The Solvay Science Project and the first Quantum Revolution

The Solvay Science Project and the First Quantum Revolution

Franklin Lambert
Alessio Rocci

Vrije Universiteit Brussel (VUB), Belgium
International Solvay Institutes, Belgium

International Solvay Institutes Brussels

World Scientific

Published by

World Scientific Publishing Co. Pte. Ltd.
5 Toh Tuck Link, Singapore 596224
USA office: 27 Warren Street, Suite 401-402, Hackensack, NJ 07601
UK office: 57 Shelton Street, Covent Garden, London WC2H 9HE

Library of Congress Control Number: 2025028363

British Library Cataloguing-in-Publication Data
A catalogue record for this book is available from the British Library.

THE SOLVAY SCIENCE PROJECT AND THE FIRST QUANTUM REVOLUTION

ISBN 978-981-98-1814-3 (hardcover)
ISBN 978-981-98-1815-0 (ebook for institutions)
ISBN 978-981-98-1816-7 (ebook for individuals)

For any available supplementary material, please visit
https://www.worldscientific.com/worldscibooks/10.1142/14447#t=suppl

Desk Editors: Nambirajan Karuppiah/Carmen Teo Bin Jie

Typeset by Stallion Press
Email: enquiries@stallionpress.com

"In no other human endeavour are
the narrow limits of nationality
and politic more obsolete and
out of place than in science."
Victor Weisskopf, 1963.

We dedicate our work to the memory of *Lars Brink* (1943–2022) and *John Lewis Heilbron* (1934–2023), two inspiring personalities, who encouraged us in our project for the safeguarding and promotion of the International Solvay Institutes archives, and who unfortunately left us during the writing of this book.

Foreword

Modern physics is often said to have started with the detection of radio waves, the discovery of the electron, and the appearance of a variety of mysterious phenomena, such as X-rays and radioactivity. Theoretical physicists may feel inclined to situate its beginnings in 1900, with Max Planck's discovery of the quantum of action. Experimental physicists may argue that quantum theory only imposed itself nine or ten years later, with the experimental verification of Planck's theory. Indeed, the actual proof of the validity of Planck's hypothesis did not arise, as one might think, from the study of black-body radiation; it was provided by the low temperature measurements of a Berlin team, headed by Walther Nernst, a physical chemist interested in the thermal properties of matter. This fact, with its historical significance, has been highlighted by Max Jammer ([Jammer (1989)], p. 45):

> If it is agreed that a physical theory becomes significant only if it also accounts for phenomena other than those for whose explanation it was conceived, then it was Einstein's paper of 1907 on Planck's theory of radiation and the theory of specific heats which endowed Planck's work with physical significance.

Nernst's verification of the predictions of Einstein's specific heat formula was an undeniable success. However, to be recognised, this success had to be brought to the attention of the elite of the profession. That happened in October 1911 at a "Physics Council" convened in Brussels by Ernest Solvay, a wealthy Belgian industrialist and scientific philanthropist. The meeting marked the recognition that classical physics was not able to account for a variety of phenomena, spreading the feeling that the classical era had come to an end. Hence, it is legitimate to associate the beginning of the so-called *quantum revolution*[1] with this international event; the more so since in its aftermath the French mathematician Henri Poincaré, demonstrated that Planck's energy quanta had to be accepted as a necessary theoretical ingredient.[Poincaré (1911)] The idea of linking the beginning of the quantum revolution to the first Physics Council finds support in the foreword to Michael Eckert's book *Establishing Quantum Physics in Munich*, written by Christoph Lehner, Jürgen Renn and Alexander Blum [Eckert (2020)]:

> In a relatively short time span following the first Solvay Conference in 1911, a discernable sub-discipline of physics emerged: quantum theory. The small group of physicists that gathered in Brussels in 1911 shared an interest in the theoretical introduction of the quantum of action which had been proposed by Max Planck in 1900 and was quickly applied to a variety of physical phenomena.

Five Physics Councils took place in Brussels under the chairmanship of Hendrik Antoon Lorentz, the greatest theoretical physicist of his time. The "Minutes" of these legendary meetings were published in French.[2] Key points discussed at the Solvay Councils have been commented on by distinguished historians of science,[3] and by the following architects of modern physics: Albert Einstein, Ernest Rutherford, Arnold Sommerfeld, Niels Bohr, Maurice de Broglie, Arthur H. Compton, Louis de Broglie, Werner Heisenberg and Paul Adrien Maurice Dirac.

As pointed out by Heisenberg, it took physicists sixteen years (1911–1927) to achieve the conceptual leap from classical to quantum physics. This happened in October 1927 at the fifth Physics Council ([Heisenberg (1929)], p. 495):

> When analysing the development of quantum theory, one must not forget in particular the discussions at the Solvay Congress in Brussels in 1927, which was chaired by Lorentz. The congress contributed extraordinarily to

[1]This term is commonly used nowadays for reasons we explain below.

[2]These *Minutes* can be consulted on line on the website of the International Solvay Institutes: `http://www.solvayinstitutes.be/html/solvayconf_physics.html`.

[3]Notably, Léon Rosenfeld, Martin Klein, Thomas Kuhn, John L. Heilbron, Abraham Pais, Max Jammer, Diana Kormos Buchwald, Helge Kragh, Bruce Wheaton, Michael Eckert.

> the clarification of the physical foundations of quantum theory through the possibility of discussion between the representatives of different research directions; it formed, to a certain extent, the completion of quantum theory.

The conceptual revolution that led to this completion happily coincided with Lorentz's presidency of the Physics Councils. Today, a technological revolution – the second quantum revolution – is gaining traction. According to Arthur Herman[4]: "It is happening now; quantum computing is going to change everything, from medicine, the study of climate change and the internet, to space exploration and how we envision national security." However, the fundamental ideas on which these technologies are based were introduced into physics almost one century ago. They are a legacy of the process that started at the first Solvay meeting (1911) and ended, as emphasised by Heisenberg, at the fifth Council (1927) with the *completion* of quantum mechanics. It must, however, be said that Heisenberg's use of the term completion reflects the general acceptance by the post-war generation of atomic physicists of the "Copenhagen view" defended in 1927 by Bohr, Heisenberg and Born. Today, we know that the meaning and scope of *entanglement*, a fundamental quantum concept with far-reaching implications, was far from being understood at the time of the fifth Solvay meeting.[5] This fact has been highlighted by Alain Aspect in the introduction of John Bell's *Speakable and unspeakable in quantum mechanics* ([Bell and Aspect (2004)], p. xviii):

> Before the realisation of the importance of Bell's theorem, which happened only in the 1970's, the conventional wisdom among physicists was that the "founding fathers" of quantum mechanics had settled all the conceptual questions ... It is my claim that Bell's example helped physicists to free themselves from the belief that the conceptual understanding that had been achieved in the 1940's was the end of the story. I think it its not an exaggeration to say that the realisation of the importance of entanglement and the clarification of the quantum description of single objects have been at the root of a second quantum revolution, and that John Bell was its prophet.

[4]A. Herman is the director of the Quantum Alliance Initiative at Hudson Institute.
[5]A first paper on the subject was published by Schrödinger in 1935.

The theorem Aspect is referring to is dated 1964. Entanglement is one of the ingredients of quantum-information science.

The present book focuses on the first quantum revolution. It aims at drawing attention to the role played in this extraordinary process by the project conceived by Ernest Solvay, which we call the *Solvay Science Project.* Its design was brought to fruition by the foundation of an International Institute, responsible for extending fundamental knowledge in physics and physical chemistry by means of two complementary actions: the support of new experiments, and the organisation of Councils similar to the one of 1911. The book pays special attention to the results obtained by researchers who benefited from the Solvay Science Project, either by taking part in one of the Councils organised by the International Solvay Institute of Physics (ISIP) during Lorentz's presidency, or through a research subsidy, granted by ISIP before the outbreak of World War I. As a result of this devastating event, our choice led to a slight imbalance in the presentation of the two approaches that "converged" in 1926, and are now an integral part of quantum mechanics. These approaches gave rise in 1927 to the various, though partially equivalent,[6] theoretical proposals discussed at the fifth Physics Council: Louis de Broglie's *pilot wave theory*, the Heisenberg–Born–Jordan *matrix mechanics*, Dirac's *transformation theory*, and Schrödinger's *wave mechanics.*

Indeed, if we map out the steps that led to these historical proposals we see two distinct lines of research, based both on Bohr's atomic theory. The first one, the *Schrödinger line*, originated in the early studies of X-rays, and focused on the successes of the Bohr model. Rooted in the idea that the description of the microscopic world should be deterministic, it involved physicists who, in one way or another, benefited from the Solvay Science Project. Their names are well known: Charles Glover Barkla, Arnold Sommerfeld, Max (von) Laue, William H. Bragg, William L. Bragg, Henry Moseley, Johannes Stark, Maurice de Broglie, Arthur H. Compton, Louis de Broglie, and Erwin Schrödinger. The line started in earnest in 1921 with Maurice de Broglie's validation of Bohr's revolutionary postulates.

[6]In 1927, before the fifth Council, Schrödinger had already given proof of the equivalence between his and Heisenberg's approach. (See [Mehra and Rechenberg (1987)], p. 636.) However, a rigorous proof of their mathematical equivalence, including a discussion of Dirac's proposal, would be given later by John von Neumann [Hove (1958)]. As for Louis de Broglie's pilot wave theory, it would become known as a step towards a deterministic interpretation of quantum mechanics [Freire (2015)].

It gave rise in 1923 to the "Compton effect" which established the long-disputed "wave-particle" nature of light. Having reached a first peak in 1926, with Schrödinger's transformation of Louis de Broglie's ideas into a mathematical formalism, it culminated in 1927 with the discovery of Louis de Broglie's matter waves, thus extending the wave-particle duality of light to the world of matter.

The second line of research, the *Heisenberg line*, originated in the perception of the failure of extending Bohr's atomic theory from hydrogen-like to generic many-electron atoms. Convinced of the need of a more radical departure from classical ideas, Hendrik A. Kramers, Wolfgang Pauli and Max Born started looking for a new mechanics that would no longer refer to Bohr's atomic orbits. This approach triumphed in July 1925 with Heisenberg's proposal of a quantum theory, formulated in terms of observable quantities, reinterpreted by Born and Pascual Jordan as a "matrix mechanics", and given a more elegant form by Dirac. The *Heisenberg line* of research did not benefit from ISIP before October 1927. However, it was Heisenberg himself who in 1929 made the already quoted statement which in the authors's eyes, justified the writing of a book. Details of the discussions to which Heisenberg refers can be found in the *Minutes* of the fifth Physics Council, published in 1928.[7] Regarding this point it should be emphasised that the volumes dedicated to the proceedings of the early Solvay meetings (1911–1927) are too often overlooked by those interested in the advances made during this extraordinary period. The little attention paid to the Solvay reports has been highlighted by Pauli, the man remembered as *the conscience of physics.* Here are the lines he wrote on January 18, 1955, in a letter to Léon Rosenfeld, one of Bohr's close collaborators:

> I still have to thank you for your letter of December 30. I think that the historians of science of the future won't have troubles with us, as all important facts are published and easily accessible ... I think that it is really unique that an author of a textbook on quantum mechanics never had even a glance at the report of the Solvay meeting of 1927. Every student here (Zurich) had heard at least the

[7]See the following books: [Solvay V (1928)] and [Bacciagaluppi and Valentini (2009)].

> word "pilot wave" connected with L. de Broglie, even if he could not know the details about it ... When I was a student myself (1918-1921), I could almost quote by heart what everybody had said on the quantum of action at the Solvay meeting of 1911 (I knew, for instance, one place where H. A. Lorentz had said "This is probably a computational error", and I also knew that actually it was not). So, you see: I always liked historical original documents (including older times), but I never liked review articles. So it remained until today.

Responding after almost seventy years to Pauli's concern about the lack of attention given to the *Minutes* of the Solvay meetings, the Unesco World Heritage Committee approved on June 8, 2023, the listing of the Archives of the Solvay Conferences in Physics and Chemistry (for the years 1910–1962) in its "Memory of the World" Register.

As regards the present work, we hope that it may be considered as a complement to the four books of a recent series *SpringerBriefs in the History of Science and Technology*, published in anticipation of the centenary of the birth of Quantum Mechanics:

Establishing Quantum Physics in Göttingen: David Hilbert, Max Born and Peter Debye in Context, 1920–1926, by Arne Schirrmacher (2019).

The Copenhagen Network. The Birth of Quantum Mechanics from a Postdoctoral Perspective, by Alexei Kojenikov (2020).

Establishing Quantum Physics in Berlin: Einstein and the Kaiser Wilhelm Institute for Physics (1917–1922), by Giuseppe Castagnetti and Hubert Goenner (2020).

Establishing Quantum Physics in Munich. Emergence of Arnold Sommerfeld's Quantum School, by Michael Eckert (2022), a book we already quoted.

Indeed, thanks to Ernest Solvay and H. A. Lorentz, Brussels became a regular meeting place for physicists and physical chemists, as well as the headquarters of an international effort aimed at developing scientific collaboration beyond national borders.

Franklin Lambert – Alessio Rocci

Introduction

Ernest Solvay (1838–1922)
Courtesy of the Solvay Institutes

History's first international conference on quanta was a "private" meeting of about twenty leading scientists, held in Brussels from October 30 to November 3, 1911. Two peculiarities of this first *Physics Council* are worthy of being recalled. Unlike the first international Congress in Physics, held in Paris in 1900, the Brussels meeting was not the work of a professional society. It was convened by an industrialist, Ernest Solvay, and took place in a country devoid of researchers likely to be interested in Planck's energy quanta. The idea of organising an *International Scientific Council* did not come from Solvay, but from Walther Nernst, the director of the Berlin Institute of Physical Chemistry. This daring chemist was the author of an ambitious, but still controversial *heat theorem.* If verified, this proposition, published in 1906, would be worth a Nobel Prize.[8] Nernst's efforts to achieve this goal, were, as we will see, at the origin of the first Solvay Council. Towards the end of 1909, measurements of specific heats at low temperature, carried out by Nernst's Berlin team, showed that Planck's quantum hypothesis could be successfully exported from the field of radiation to the field of matter.

Einstein's role in Nernst's conversion to quantum theory has been discussed elsewhere.[9] Let us just mention that after meeting Einstein in March 1910, Nernst became convinced that his theorem could be deduced from Einstein's quantum theory of specific heats. Galvanised by the idea, he looked for a way to have Planck's hypothesis validated by leading physicists. He soon realised that this required the convening an *International Scientific Council.*

But where should that Council take place? It was imperative that it be organised on "neutral ground" by an independent, preferably non-German, patron of science.[10] Nernst immediately thought of Ernest Solvay, the Belgian industrialist, passionate about physics, who had recently come to Berlin to receive the Leibniz Medal of the Prussian Academy of Science (1909).[11]

---

[8]Reformulated by Planck, the heat theorem would become known as the third law of thermodynamics. However, at the time of its formulation, Nernst's proposition was still far from being accepted. (in fact, his Nobel Prize was delayed for about fifteen years: see [The Nobel Foundation (1962)], p. 389).

[9]See [Lambert and Berends (2021)], pp. 30–40.

[10]Nernst, who aspired to a Nobel Prize for his theorem, knew that he had to reckon with the firm opposition of his Swedish rival, Svante Arrhenius. He could not afford to appear as the initiator of an operation aimed at strengthening his chances.

[11]This distinction had been granted to him on the proposal of Nernst and Emil Fischer.

Things happened as Nernst hoped: Solvay accepted, at least in principle, to convene the *Council* in Brussels. Having his own reasons to consider physics a science in crisis, the industrialist got involved[12] in the operation towards the end of June 1910. The timing of the *Council* proved crucial. Ten years had passed since Planck's discovery of the quantum of action h. Yet, his theory, based on the assumption of *discontinuous* energy exchanges between matter and radiation (quanta), was still far from being accepted. Nernst therefore saw the *Council* as a means of publicising the success of Planck's theory in the molecular domain. Experience proved him right. Not only did the meeting establish a consensus on the fact that "discontinuities in energy" could no longer be ignored, it also brought Planck's quanta to the attention of a broad community of physicists and physical chemists.

For Solvay, the real benefit of the Council was his meeting with Hendrik A. Lorentz, the great Dutch master of theoretical physics, whose conduct of the debates had been admired by everyone. Here was a man who could be called on to carry out a project close to Solvay's heart: the founding in Brussels of an *International Physics Institute.*

The industrialist's master plan – the Solvay Science Project – was twofold. It aimed, on the one hand, at stimulating progress in physics and physical chemistry at the international level. On the other, it could mitigate the effects of the glaring lack of qualified physicists in Belgium. To achieve these goals, Solvay thought of creating two International Institutes. Both would be charged with a triple mission: to grant subsidies to promising scientists from everywhere, to organise regular "summits" in physics and chemistry, to grant scholarships to Belgian researchers wishing to improve their skills in foreign laboratories.

This book deals with the aspects of the *Solvay Science Project* relating to physics and physical chemistry.[13] Limiting itself to discussing the Physics Project's impact on the quantum revolution, it first recalls two events: the convening in June 1911 of the first Physics Council, henceforth Solvay I, and the subsequent founding of the *International Solvay Institute of Physics* (ISIP). It then focuses on the actions carried out by the Institute before

[12]Solvay met Nernst during the early summer of 1910. He realised that the Council would provide him with the opportunity to present his "gravito-materialitic" theory to an authorised body. Thrilled by the idea, he asked his collaborators to examine Nernst's proposal, but waited until March 1911 before giving his final answer (see [Lambert and Berends (2021)], section 1.3, for details).

[13]The Solvay Chemistry Councils started in 1922. A historical account of their impact is still lacking.

the outbreak of the Great War, and on its post-war activities during the second period of Lorentz's presidency.[14]

Regarding the first Physics Council, it is clear that Solvay intervened in the drawing up of the list of participants. His goal was to maintain an equal number of invitees representing France, Germany and the United Kingdom. Thus, it was thanks to Solvay that Marcel Brillouin, Marie Skłodowska Curie, Maurice de Broglie and Henri Poincaré attended the meeting. Their presence would be rich in consequences.

In his invitation of June 1911, Solvay specified the Council's purpose. It was the one that had been set by Nernst: "To elucidate current questions of molecular and kinetic theory". However, Solvay I (1911) went down in history as a conference on "The theory of radiation and quanta".[15] This change, reflected by the title of the volume published in 1912, was due to Lorentz's decision to highlight the fact that the debates gradually shifted from the thermal properties of matter to the more fundamental problem of radiation.

The founding of ISIP, an idea reinforced by the Council's success, materialised on May 1, 1912, thanks to Lorentz's proposal for an Institute that would meet the founder's objectives. A key-element in ISIP's design was the decision to entrust its management to an independent "*International Scientific Committee*" (ISC). Financial and administrative matters would be the responsibility of a local "Administrative Commission" (AC). This "two-headed" management had been suggested by Wilhelm Ostwald in an earlier idea for an Institute of Chemistry, submitted to Solvay in April 1911. Ostwald's project aimed at the establishment of an International Chemistry Institute in partnership with the *International Association of Chemical Societies* (IACS).

As for ISIP's main actions, the organisation of Physics Councils and the granting of research subsidies, they led during the years 1912–1914 to the convening in October of Solvay II (1913), and the support of about 40 research projects.

A second part of the book is devoted to ISIP's reorganisation after the First World War. It focuses on the convening of the three Physics Councils that took place under Lorentz's presidency during the years 1921–1927. The Great War and its consequences, the collapse of international relationships and the economical difficulties of the belligerent countries,

---

[14]In this introduction, the various Physics Councils are followed by the year they took place in. In the rest of the book, when quoting the published volumes, we shall return to the usual quotation habit by indicating the publishing year, e.g. [Solvay I (1912)].

[15]The French title reads: "La théorie du rayonnement et les quanta".

heavily affected the "Solvay Project". As a result of the depreciation of the Belgian franc, ISIP could no longer support the granting of subsidies for the purchase of scientific equipment. The post-war task of the International Scientific Committee was reduced to the organisation of Physics Councils. Another consequence was the temporarily exclusion of German scientists, a decision made inevitable by the fact that many of them, including Planck and Nernst, had signed the *Appeal to the Civilised World*, an outrageous pamphlet of October 1914 that denied the violation of Belgian neutrality. This exclusion meant that major advances in quantum theory could not be validly discussed during the first two post-war Councils (Solvay III and IV). However, as we will see, all Solvay meetings organised under Lorentz's presidency deserve special attention. Indeed, the writing of this book was motivated by the observation that the *Solvay Science Project*, and its achievements in the field of physics, thanks to ISIP's actions from the spring 1912 to the fall of 1927, had a direct and significant impact on the quantum revolution.

First, we specify what we understand by "quantum revolution". As indicated in our foreword, we claim that it was a process stimulated by the convening of the first Council in 1911, which started with Poincaré's proof that the introduction of Planck's hypothesis could not be circumvented [Poincaré (1911)]), and ended sixteen years later with the establishment of a synthesis between the two "physical worlds" long considered fundamentally distinct: the world of electromagnetic radiations, described in terms of continuous waves, and that of ponderable matter, described in terms of *discrete* atoms and molecules.[16] This synthesis resulted to a large extent from X-ray research carried out by first-rank experimentalists, such as Walter Friedrich, Max (von) Laue, William Henry Bragg, Henry Moseley, Maurice de Broglie, Arthur Holly Compton. It took shape thanks to the ideas put forward by leading theoreticians: Albert Einstein, Niels Bohr, Arnold Sommerfeld, William Lawrence Bragg, Louis de Broglie, Wolfgang Pauli, Werner Heisenberg, Max Born, Pascual Jordan, Erwin Schrödinger, and Paul Adrien Maurice Dirac. Our point is that the quantum revolution – a process triggered in 1900 by Planck's discovery of the universal constant h representing the quantum of action – started in earnest in the aftermath of Solvay I, and culminated in 1927, at Solvay V, with the *completion* of Quantum Mechanics.[17]

[16]It was this lack of harmony between the descriptions of these two aspects of physical reality that prompted Einstein to look into the problem of black-body radiation.

[17]In the Foreword, we already specified the meaning of the word *completion* used by Heisenberg.

Three statements support this view. One, about the end of the process, is Heisenberg's statement of 1929 that we quoted in the foreword to this book. Another, about the start of the process, is the excerpt of the preface to Eckert's book of 2022, also quoted in our foreword. A third statement, about the impact of the discussions at Solvay I, was made by Marcel Brillouin in a letter to Lorentz of January 30, 1912. Reacting to Poincaré's paper of December 1911, he told his Dutch colleague[18]: "Did you read Mr. Poincaré's penetrating note? Would Mr. Solvay have caused only this result, he would not have wasted his time."

Indeed, by settling the question of the status of the quantum hypothesis (a problem that the Council had not been able to resolve), Poincaré's proof of its inescapability, marked the beginning of the quantum era.[19] The French master's authority was such that his conclusion imposed itself on James Jeans, a British member of Solvay I, who had opposed the introduction of quanta before becoming an ardent defender of Planck's theory.[20]

Let us now point at two remarkable twists of fate. Firstly, we were struck by the fact that the sixteen years necessary to achieve the conceptual leap from classical to quantum physics coincided with a *golden age* during which ISIP's direction was placed under the enlightened chairmanship of H. A. Lorentz.

Secondly, as we focused on Arthur Compton's "discovery" of the photon, we were struck by the fact that it took the same sixteen years for the Nobel Committee for Physics to recognise the role played by Charles Thomson Rees Wilson's *cloud chamber*, a device developed in 1911 which allowed the observation and photography of the traces produced in a gas by the passage of ionising particles.[21]

In fact, there is a third point that strengthened us in our intention to take stock of the developments to which the *Solvay Science Project* made an effective contribution. Indeed, many articles and books provide detailed information on the five Solvay meetings that took place during ISIP's "golden age" and on the way in which they contributed to the

---

[18]Letter from Brillouin to Lorentz, Haarlem, Noord-Hollands Archief (NHA), Archief H. A. Lorentz (364), inventaris nummer 11.

[19]See [McCormmach (1967)], p. 39. As pointed out by this well-informed author: "Poincaré's response was the most striking single repercussion of the Congress."

[20]See [McCormmach (1967)], p. 42.

[21]See [The Nobel Foundation (1962)], p. 488.

development of new research areas in physics (X-ray spectroscopy, atomic physics, solid state physics).[22] However, despite these publications, little effort has been made so far to collect the relevant pieces of information and bring them together into a comprehensive account. More importantly, there is almost no trace in the literature[23] of the subsidies granted by ISIP before the outbreak of World War I, nor does the composition of the first International Scientific Committee appear in the published reports of the pre-war Council.[24] This lack of attention for the Institute's subsidy programme[25] is all the more regrettable since it supported X-ray research, one of the pillars of quantum theory that led to the understanding of the structure of atoms, the discovery of the dual "wave-particle" nature of light and matter, the understanding of crystal structure, and the establishment of the laws that govern the properties of chemical elements. About half a dozen Nobel Prizes were awarded for work supported by Solvay grants[26] during the years 1912–1914, a period described by Robert Millikan as "being comparable in importance with the one which saw the birth of Galilean-Newtonian mechanics some three centuries earlier."[27]

It is the purpose of this book to fill the above gaps, or at least to reduce their extent, by showing that ISIP's action had a clear impact on the development and completion of the (first) quantum revolution.

Two elements forced us to divide our story into two parts: the dimension of the challenge, and the changes in the Institute's organisation caused by the Great War. As anticipated, part I is devoted to the pre-war period. It focuses on the two "complementary" tasks entrusted to the Physics

[22] We should mention, in particular, Niels Bohr's report "The Solvay Meetings and the Development of Quantum Physics", presented in 1961 at the twelfth Physics Council (see [Solvay XII (1962)], pp. 13–36).

[23] A noteworthy exception is Heilbron's report of a subsidy granted in 1913 to Henry Moseley which enabled him to continue his work (see [Heilbron (1966)]).

[24] At the time of ISIP's foundation (May 1912), the ISC comprised nine members: H. A. Lorentz, Marie Skłodowska Curie, M. Brillouin, W. Nernst, Emil Warburg, E. Rutherford, Heike Kamerlingh Onnes, Martin Knudsen, Robert Goldschmidt. However, due to the war, ISIP's Statutes were only published in 1920, i.e. after the replacement of Nernst and Warburg (Germany's representatives) by Augusto Righi (Italy) and William Henry Bragg (United Kingdom).

[25] ISIP's subsidy programme is only briefly mentioned in (Mehra, 1975) and (Marage and Wallenborn, 1999).

[26] Max (von) Laue (subsidy in 1912, Nobel Prize in 1914), Charles G. Barkla (subsidy in 1912, Nobel Prize in 1917), James Franck and Gustav Hertz (subsidy in 1913, Nobel Prize inn 1925), William L. Bragg (subsidy in 1914, Nobel Prize in 1915), Johannes Stark (subsidy in 1914, Nobel Prize in 1919).

[27] See [Kragh (2010)], p. 3.

Institute: the organisation of a second Council (Solvay II) and the support of experiments on topics highlighted by the first two Councils.

We will see that from July 1912 to June 1914, the main concern of the International Scientific Committee was to evaluate the experimental projects submitted to the Institute. Its other task was to bring together a limited number of experts in order to analyse all aspects of the chosen topic for Solvay II, The Structure of Matter, and to set new research directions. Special attention is given to the way in which reports, presented at Solvay I and II, influenced decisions taken by the ISC with regard to the granting of subsidies. Among the hundred projects submitted to ISIP before the start of the Great War, around 40 were considered worthy of support. More than ten physicists, named by Bohr in his *Rutherford Memorial Lecture* [Bohr (1961)], benefited from a Solvay grant. Some are famous: Sommerfeld, Barkla, Laue, W. L. Bragg, Franck, Hertz, Moseley, Stark. Other lesser-known researchers are cited by Bohr for a specific contribution: Robert Wood, Louis Dunoyer, and Kasimir Fajans.

The complete list of the supported projects shows that an important part of the available resources was used for the purchase of X-ray equipment.[28] It tells us that Lorentz and his colleagues were aware of the new insights that X-rays could provide in the new areas already mentioned. Particularly striking are the disinterested efforts of major scientists, such as Rutherford and Marie Skłodowska Curie, who agreed to spend part of their precious time in evaluating the numerous projects that flowed into the Institute before the outbreak of World War I.[29] Their commitment to ISIP is a moving testimony of their faith in the universal role of science, and in its value as a contributor to peace. Unfortunately, this faith was annihilated in August 1914 by the outbreak of the First World War.

Part II of the book is, as we said, devoted to ISIP's post-war reorganisation, and to the Councils that took place during the period 1921–1927. The Great War could have caused the end of Solvay's Physics Institute. The danger of a dissolution was real, but ISIP survived the collapse of international relations thanks to the conjunction of three factors: Lorentz's conciliatory attitude, the understanding shown by Solvay, and ISIP's statutory provisions. Unfortunately, the absence of German physicists at Solvay III (1921) and Solvay IV (1924) produced the imbalance mentioned in the foreword: the early stages of the *Heisenberg line*, mostly

[28]See [Lambert and Berends (2021)], pp. 168–170.

[29]Rutherford, known for his reluctance to write reports and grant proposals, wrote no less than eight reports on projects that were submitted to ISIP before October 30, 1913.

developed by German researchers, were not discussed at the firsts post-war Councils. Thanks to a proposal made by Lorentz, and approved by the Belgian King, German physicists were invited in 1927 to attend Solvay V (1927).[30]

Solvay III and Solvay V are specially highlighted as landmarks in the verification of Bohr's atomic theory, and the transition from the "Old Quantum Theory" to Quantum Mechanics. Solvay IV, on the other hand, stands out as the Council attended by Schrödinger before realising the work that would make him famous. It was at this meeting, also attended by Peter Debye and Abram Fedorovitch Joffé (the "father" of Soviet physics), that Langevin mentioned Louis de Broglie's thesis in private, and praised its originality. Thanks to Joffé's reminiscences of Solvay IV, published in 1967, we have first-hand information on little-known aspects of Solvay IV: Joffé's contacts with Peter Debye, Langevin and Schrödinger, and his meeting with King Albert.[31] Some of them are reported in Appendix A.

Looking back at the five Councils organised under Lorentz's presidency, it appears that most of them focused, at least initially, on the structure and the properties of matter. This was clearly the case for Solvay I, intended to elucidate problems of kinetic and molecular theories, Solvay II, entitled *On the structure of matter*,[32] Solvay III, which focused on *atoms and electrons*, and Solvay IV, dedicated to the electric conductibility of metals. However, during these Councils, attention was also paid to the role of radiation. At Solvay I several reports were devoted, in part, to the interaction between matter and radiation. At Solvay II, X-rays were identified as periodic electromagnetic radiations similar to light; at Solvay III they were shown to provide key-information on Bohr's description of the structure of atoms. At Solvay IV, discussions repeatedly focused on the role of X-rays[33] in the analysis of crystal structure. However, the dual nature of light emerged gradually. Solvay V, entitled *Electrons and Photons*, became the Council at which de Broglie's *matter waves*, and Bohr's *complementarity principle* were highlighted as revealing elements of the fundamental unity of matter and radiation.

Before concluding this introduction and beginning our journey through the Solvay Councils, let us return to the founding project which enabled these meetings to take place, and to perpetuate the *Physics* and *Chemistry*

---

[30] This action can be interpreted as an act of "scientific diplomacy".

[31] See Part II, chapter 5.

[32] The French title reads: "La structure de la matière".

[33] Their role with regard to the acceptance of Einstein's light quantum hypothesis was emphasised by Joffé in his report on "The Propagation of Radiation Pulses".

Councils to this day. From a more general perspective, the *Solvay Science Project* may be seen as an initiative in support of Angelo Baracca's claim that the beginning of the twentieth century was a period during which "the organisation of scientific research evolved from one of individual scientists, investigating on the basis of their personal resources and interests, to a complex structure, financed by the State or by corporations".[34] In this sense we can say that the *Solvay Science Project* is representative of an "intermediate period" during which scientific research was supported by private funds made available by industrialists and bankers.[35]

However, the action carried out by Solvay stands out for several reasons. One is the fact that this autodidact had naively set himself the goal of constructing, from reflection alone, an "all-encompassing theory" of the Universe. Another is that in 1910, when he resumed his investigations, Solvay was aware, for personal reasons, that physics was in crisis. A third reason for action was the lack in Belgium of well-trained physicists, a situation which prompted Solvay to call on foreign experimenters and set up a system of grants to support their research.

Last but not least, the success of the Solvay Science Project was intimately linked to the industrialist's decision not to interfere with the Institute's management, nor with the priorities set by the *International Scientific Committee* in the granting of subsidies (a decision all the more commendable since Solvay ardently wished to promote new experiments likely to confirm the predictions of his "Gravito-Materialitic" theory).

The book relies heavily on analyses made by first rank historians of science, such as those cited in the foreword. Excerpts from their books and papers are quoted whenever deemed useful. We also quote a variety of statements made by major contributors to the quantum revolution. These include Planck, Einstein, Rutherford, Bohr, Maurice de Broglie, Siegbahn, Louis de Broglie, Heisenberg, Compton and Dirac. The order in which the successive developments are presented reflects, as far as possible, the chronological order of events.

We also included four appendices. The first one is devoted to Joffé's reminiscences of Solvay IV. The other appendices (B, C, and D) provide additional information on two subjects: Bohr's long-standing refusal to accept Einstein's light quanta, and the recognition by the Nobel Committee

[34]See [Baracca (2005)], p. 50.

[35]Apart from Solvay and well-known Americans, such as Andrew Carnegie, John D. Rockefeller, and Henry Goldman, one may cite Germans, such as Geheimer Kommerzienrat Lepold Koppel, Franz Oppenheim, director of the Aktiengesellschaft für Anilinfabrikation, and the Recklinghausen industrialist Carl Still.

of the role played by Wilson's cloud chamber (a device referred to throughout the book) in several key-stages of the quantum revolution.

Detailed information on the actual steps that led to the formulation of quantum mechanics can be found in reference works, e.g. in Max Jammer's book [Jammer (1989)].

The reader wishing to obtain more details on the early Physics Councils should consult the sources included in the bibliography, and in particular the reports of the first five Councils, edited in French and published in Paris by Gauthier–Villars.

The Archives of the Solvay Institutes which are kept by the *Service des Archives et Bibliothèques de l'Université Libre de Bruxelles* (henceforth *S.a.b. ULB* in our quotation of the original documents), contain a collection of original Solvay reports. Credits for the pictures digitized by *S.a.b. ULB* and contained in this book are given through the statement *Courtesy of the Solvay Institutes.*

A report of Solvay I, written in German, and edited by Arnold Eucken, was published in 1914 by the German editor Wilhelm Knapp Verlag in the *Abhandlungen der Deutsche Bunsen-Gesellschaft* under the title: *Die Theorie der Strahlung und der Quanten: Verhandlungen auf einer von E. Solvay einberufenen Zusammenkunst (30 Oktober bis 3 November 1911).*

An English translation of the Minutes of Solvay V has been provided by Guido Bacciagaluppi and Anthony Valentini in their book entitled *Quantum Theory at the Crossroads: Reconsidering the 1927 Solvay Conference* [Bacciagaluppi and Valentini (2009)]. As regards the Minutes of the Solvay meetings for the period 1913–1924, we provide a number of original quotations in English (taken from the Solvay archives), as well as translated versions of excerpts from the volumes published in French.

Contents

PART I

Early Solvay Councils (1911–1913) and Subsidy Programme of the International Physics Institute (ISIP)

Chapter I.1

Solvay I (1911): The theory of radiation and quanta

Solvay I – October 30 to November 3, 1911
Courtesy of the Solvay Institutes

I.1.1 From Planck's radiation law to Solvay I

Physics students are usually told that the beginning of the quantum era was signalled by Planck on December 14, 1900, when he presented his derivation of the distribution law for *black-body radiation* to the German Physical Society. In an attempt to justify a formula that accounted for the "black-body" data in all regions of the spectrum, Planck assumed that the exchange of energy between matter and radiation was a *discontinuous* process, taking place by indivisible units. To fix the magnitude of these energy units (quanta), he introduced a constant h – *the quantum of action* – foreign to all existing theories. As noticed by Luis J. Boya, the symbol h stands for "helping quantity" (HilfGrösse) ([Boya (2003)], p. 92), a term that clarifies Planck's interpretation of the new constant.[1]

In 1905, Einstein examined the black-body problem independently of Planck's work. He noticed that the application of classical concepts resulted in an absurdity: the presence of an infinite amount of radiant energy in a finite volume. Forced to abandon classical theories, Einstein focused on the source of the absurdity: the high frequency radiations, which happened to obey Wilhelm Wien's radiation law. After a thorough examination of that law, he came to the conclusion that radiations of high frequency behave as a "gas of energy quanta" [Einstein (1905)]. Einstein thus shifted the source of the problem from the interaction between matter and radiation to the very nature of light. To test his "heuristic" point of view, which completely deviated from the accepted wave theory, Einstein applied it to the photoelectric effect (the emission of electrons by a metal irradiated with ultraviolet light). He showed that his "light quantum hypothesis" provided a simple explanation of the puzzling aspects of the effect. However, the reality of "photons" (i.e. light quanta endowed with a momentum) would only be recognised twenty years later with the general acceptance of Compton's results for scattered X-rays, i.e. ten years after Millikan's validation of Einstein's photoelectric equation.

On the other hand, the first verification of Planck's quantum theory, took place towards the end of 1909, with Nernst's validation of Einstein's specific heat formula of 1907, a first success of the theory recorded in the molecular field ([Jammer (1989)], p. 45). Details on the sequence of events that led to this major result have been reported elsewhere.[2] Here, we will limit ourselves to recalling some key points.

---

[1] Planck, at the time, did not realise the significance and scope of his discovery.

[2] See [Lambert and Berends (2021)], pp. 35–45.

It all started in 1906, with the publication by Nernst of an ambitious *heat theorem*. This controversial proposition allowed the calculation of the equilibrium conditions of a chemical reaction by means of thermal data. If confirmed, it would be worth a Nobel Prize. The theorem predicted a decrease of all specific heats with temperature, and their convergence towards one and the same limit in the approach of absolute zero. To check this prediction, Nernst engaged his Berlin team in a campaign of specific heat measurements. He also looked for a theory that might support his theorem. Having been informed of Einstein's specific heat formula, the chemist decided to test its predictions at very low temperatures, down to the boiling point of air. The test proved conclusive: the measured specific heat values were found to be in line with the predictions of the formula. Enchanted by this success, Nernst went to Zurich to meet Einstein.[3] The meeting took place in the beginning of March 1910. Einstein informed his visitor of the quantum basis of his formula. He also urged him to push his measurements down to the boiling point of hydrogen, explaining that a complete verification of the formula would provide decisive proof of the validity of Planck's quantum hypothesis.

Einstein's revelations put Nernst in an awkward position. On the one hand, as Jammer observes ([Jammer (1989)], p. 47), Nernst felt reassured by the idea that his theorem could be derived from Einstein's specific heat theory. On the other, he was aware of the risk he was taking by linking the fate of his theorem to that of Planck's theory, an "incongruous" proposition that could at best be described "as a simple interpolation formula" [Jammer (1989)]. Last but not least, people in Berlin knew that Planck had been denied a Nobel Prize in 1908, on the ground that his theory was based on a "completely new and barely plausible hypothesis, namely the existence of an elementary quantum of energy"[4] ([Crawford (1987)], p. 133). Hence, there was only one way to solve the dilemma. Nernst had to take two steps. One was to redirect the quantum debate to the molecular field, where Planck's hypothesis had obtained significant support. The other was to inform leading physicists of the success of Einstein's quantum theory. Both goals required bold action: the convening of an *International Scientific Council*.

[3]This former clerk at the Swiss Federal Office for Intellectual Property in Bern had recently become an associate professor at the University of Zurich.

[4]Having been told that he would get the 1908 Physics Nobel Prize, Planck gave interviews which were published in newspapers.

The idea seemed appealing, but where should the Council take place? Its natural location was Berlin, cradle of quantum theory and seat of Nernst's Institute of Physical Chemistry. However, for reasons linked to the chemist's personal situation, it was clear that the meeting should be convened on *neutral* ground by a non-German personality, foreign to his professional sphere.[5] Brussels, the city where a World Exhibition was about to open (and a series of international congresses had been scheduled) seemed perfect. It was the home-town of Solvay, a scientific patron passionate about physics, whom Nernst had met a few months earlier in Berlin. Furthermore, it was the town in which Goldschmidt, Nernst's Belgian collaborator, and a friend of Solvay, had his headquarters. This meant that Nernst had a man in Brussels who would be able to take charge of all matters regarding the organisation of the Council.

A meeting between Nernst and Solvay was hastily arranged at Goldschmidt's home. We have no record of what was said during that meeting, but it is clear from later correspondence that Nernst refrained from revealing his plan, limiting himself to sounding out Solvay on his willingness to take part in an operation presented as "a way to solve a major crisis in physics" [Lambert and Berends (2021)].

Reassured by the industrialist's reaction, Nernst decided to send him two letters. Both were dated July 26, 1910. One was a personal message, in which Nernst asked Solvay to convene an *International Scientific Council.* He explained that he had spoken about the project to Lorentz, Planck, and Knudsen, without specifying date and location of the meeting. The Council members would be John William Strutt (Lord) Rayleigh, Lorentz, Planck, Knudsen, Einstein, Wien, Wilhelm Röntgen, Hugo von Seeliger, Joseph J. Thomson, Arthur Schuster, Jeans, Joseph Larmor,[6] Rutherford, Jean Perrin, Paul Langevin, Johannes D. van der Waals, and Friedrich Hasenöhrl.

The second letter contained the draft of a confidential invitation that Solvay was kindly asked to sign in case of agreement. It was entitled "Invitation to an International Scientific Council to Elucidate Some Current Questions of Kinetic Theory" ([Lambert and Berends (2021)]). Solvay agreed on August 5, 1910, reckoning that the Council would allow him to

[5]Nernst had a long-standing conflict with Arrhenius, an influential member of the Nobel Committee for Physics, and one of the few Swedish scientists whose opinion was authoritative in the Committee for Chemistry (see [Crawford (1987)], p. 99).

[6]Larmor was invited, but like three other British invitees, he did not participate in the first Council.

present his *Gravito-Materialitic theory*, a work to which he had just given a new impetus, but which needed to be completed. He therefore asked that the Council be postponed to the fall of 1911.

During the following months the list of invitees underwent several changes, partly to respond to the industrialist's wish to ensure a balance between representatives of major nations. Six scientists were added to those indicated by Nernst: Sommerfeld, Kamerlingh Onnes, Marcel Brillouin, Marie Curie, Poincaré and Maurice de Broglie. Two German experts on black-body radiation, Warburg and Heinrich Rubens, were chosen to replace Röntgen and von Seeliger.

Regarding the Council's presidency, it was Lorentz, the undisputed master of theoretical physics, who after two unsuccessful attempts – first Lord Rayleigh, then Planck – was asked to chair the debates.[7] This choice proved particularly judicious, notably because of the high scientific and moral qualities of the Dutch master, but also thanks to his perfect knowledge of French, German and English.

I.1.2 The first Physics Council

For reasons put forward by Heilbron [Heilbron (2015)], only two British physicists (Rutherford and Jeans) accepted to attend the Council. The others declined the invitation for various reasons.[8] A few days before the meeting, Fredrick Lindemann, Nernst's British collaborator, was invited to join Maurice de Broglie as a Scientific Secretary.

The letters of invitation, signed *Ernest Solvay*, were sent on June 15, 1911. The Council was expected to take place from October 30 to November 3, 1911. Its purpose was, as proposed by Nernst, "To Elucidate Current Questions of Kinetic and Molecular Theories".[9] [Lambert and Berends (2021)] Lorentz opened the conference with this grim picture of the situation: "We have the feeling that we are at a dead end, as the old theories are showing themselves powerless to pierce the darkness which surrounds us at all sides" [Solvay I (1912)].

[7]Nernst's idea of asking Lord Rayleigh to chair the meeting was abandoned when it became clear that he would not come to Brussels. Edouard Herzen, one of Solvay's close collaborators, proposed Planck, who apparently refused, suggesting Lorentz.

[8]Lord Rayleigh, who was among the invitees, declined the invitation to the Council, but said he would send a letter "out of respect for his colleagues".

[9]The kinetic theory of matter sought to explain the phenomena of heat in terms of the motions of the molecules.

However, seeking to reassure is colleagues, the chairman mentioned some positive elements, and traced a possible trajectory for the future [Solvay I (1912)]:

> In this state of affairs, the beautiful hypothesis of energy elements first put forward by Mr. Planck and applied to many phenomena by Mr. Einstein and Mr. Nernst, has been a precious ray of light. It has opened up unexpected perspectives, and even those who view it with a certain mistrust must recognise its importance and fruitfulness ... However, this new idea, beautiful as it is, also raises a number of objections.
>
> I won't reproach it for putting us in contradiction with the ideas on motion and the action of forces that have prevailed for centuries; I fully understand that we have no right to believe that in physical theories of the future everything will be in accordance with the rules of classical Mechanics ... It is now easy to sketch out the program we need to follow.
>
> First, we need to make clear the imperfections of the old theories, specifying as far as possible the nature and causes of their defects. We will then examine this idea of energy units in the various forms it has been given ... Finally, we would be very happy if we could get a little closer to this future Mechanics.

With these statements, Lorentz endorsed the idea that the (first) quantum revolution began in earnest at Solvay I. This idea, that we put forward in this book, has also been defended by the historian of science Dieter Hoffmann.[10]

I.1.3 Council sessions

Lorentz recalled that 11 reports had been scheduled to guide and facilitate discussions. According to Maurice de Broglie's notes, taken on the spot, the Council sessions were organised as follows [M. de Broglie (1911)]:

[10]See [Hoffmann (1984)], p. 17.

Monday – October 30, 1911

Morning session, hotel Metropole (chair: Poincaré) Report by Lorentz: "Application of the equipartition of energy theorem to radiation;" discussion.

Afternoon session, hotel Metropole (chair: Lorentz) Presentation by Lindemann of a letter from Lord Rayleigh; discussion. Report by Jeans: "The kinetic theory of specific heats according to Clausius, Maxwell and Boltzmann;" discussion.

Tuesday – October 31, 1911

Morning session, hotel Metropole (chair: Lorentz) Report by Warburg: "Experimental verification of Planck's formula in the region of high frequencies"; discussion. Report by Rubens: "Verification of Planck's radiation formula in the long wave region;" discussion.

Afternoon session, hotel Metropole (chair: Lorentz) Report by Planck: "The black-body radiation law and the hypothesis of elementary action quantities;" discussion.

Wednesday – November 1, 1911

Morning session, hotel Metropole (chair: Lorentz) Report by Knudsen: "Kinetic theory and the experimental properties of ideal gases"; discussion. *Afternoon session*, Solvay Institute of Physiology (chair: Lorentz) Report by Perrin: "Evidence of molecular reality, special study of emulsions;" discussion. First part of Nernst's report: "Application of the theory of quanta to a number of physio-chemical and chemical problems."

Thursday – November 2, 1911

Morning session, hotel Metropole (chair: Lorentz) Discussion of Perrin's report. Second part of Nernst's report; initial discussion. Presentation by Kamerlingh Onnes of electrical resistance curves for various metals.

Afternoon session, hotel Metropole (chair: Lorentz) Further discussion of Nernst's report. Report by

Sommerfeld: "Application of the action element theory to non-periodic molecular phenomena;" discussion.

Friday – November 3, 1911

Morning session, hotel Metropole (chair: Lorentz) Report by Langevin: "Kinetic theory of magnetism, the magneton;" discussion. First part of Einstein's report: "The current state of the specific heat problem." **Private meeting of Solvay with the members of a *Select Committee*.**

Afternoon session, hotel Metropole (chair: Lorentz) Continuation of Einstein's report; discussion. General discussion and closing speeches.

Most of the presentations, notably those by Planck, Nernst, Sommerfeld and Einstein, gave rise to lively discussions. All members spoke in the language of their choice. Lorentz translated questions and answers; he also summarised the debates at the end of each session. Maurice de Broglie took note of the interventions made in French, and collected handwritten notes from members who had made their remarks in German or in English.[11] It was agreed that the Council Minutes would be published in French, out of respect for Mr. Solvay.

André Langevin, son of Paul Langevin, tells us that his father was designated by Lorentz as *adjunct secretary of the Council* ([Langevin (1966)], p. 6). Invested in this role, Paul Langevin took care of the necessary translations.

The volume edited by Langevin and Maurice de Broglie [Solvay I (1912)] was published in Paris in 1912. It appeared under the title *The Theory of Radiation and Quanta*. As indicated in the introduction, this choice, which differed from the Council's initial purpose,[12] was due to a shift in focus that occurred during the meeting. The current questions of molecular theory gradually gave way to the problem of radiation. Two elements may explain this shift:

[11]Thanks to his work, we are aware of significant events that are not reported in the published Council Minutes. We also know that part of the discussions took place during the reports, an informality typical of today's seminars, favoured by the limited number of participants and the confidential character of the meeting.

[12]In French: "Conseil pour élucider quelques questions d'actualité des theories cinétiques et moléculaires".

(i) Einstein's decision to go beyond the subject of his report, and to address a problem that haunted him since his "revolutionary" paper of March 1905: the need to deal with two apparently irreconcilable properties of light.
(ii) Lorentz's particular interest in any theoretical argument that could explain the puzzling fact that, for large wavelengths, Planck's radiation law seemed to agree with his own result of 1903.[13]

I.1.4 Noteworthy presentations and events

Highlights of Solvay I have been recorded by Bohr in his paper entitled *The Solvay Meetings and the Development of Quantum Physics* and presented in 1961 at the twelfth Physics Council [Solvay XII (1962)]. Aspects of the meeting have also been discussed in a variety of papers and books,[14] including an unpublished manuscript written by Jean Pelseneer.[15] In this section, we limit ourselves to reporting a series of facts that shed light on the problem of quanta, as it presented itself ten years after the discovery of the quantum of action. We pay attention to specific issues, such as the early statements about the apparent particle-like distribution of radiant energy, the reactions to the first quantum atomic model suggested by Arthur Erich Haas, and Kamerlingh Onnes's presentation of a graph showing for the first time the phenomenon of superconductivity. We also report Einstein's comments on the scope of the problems raised by Planck's radiation formula, and his statements on the need to question the validity of two pillars of classical theory: mechanics and electrodynamics. Other noteworthy elements are Rutherford's silence about the α-scattering results that had been obtained in Manchester, and Brillouin's description of the Council members' general feeling at the end of the meeting.

We also highlight two little-known facts: the withdrawal from the Council's official Minutes of a crucial remark made by Planck, and the holding on November 3, 1911 of a private meeting of Solvay with the members of a *Select Committee*. As a complement to these points, we

[13]See [Lorentz (1934–1939)], Volume 3, p. 155.

[14]We quote only some of them in alphabetical order: [Barkan (1993); McCormmach (1967); M. de Broglie (1951); Klein (1965); Klein *et al.* (1994a); Marage and Wallenborn (1999); Mehra (1975); Rosenfeld (1936); Wheaton (1983)]. See also Heilbron in [Solvay XXV (2013)] and Peter Galison in [Solvay XXIII (2007)].

[15]See [Pelseneer (unpublished)]. This manuscript contains a copy of Planck's response – dated June 11, 1910 – to the letter in which Nernst announced his idea of convening an "International Scientific Council" to discuss the quantum problem.

mention the impact of the Brussels discussions on two future architects of quantum theory: Bohr and Louis de Broglie.

I.1.4.1 *Planck and the quantum of action*

In his report, Planck exposed an alternative to his initial theory. Based on a new quantum hypothesis, this "emission theory" appeared to him as a means to justify his radiation law, while avoiding contradictions[16] with the laws of classical electrodynamics. Planck's new theory implied the existence of an extra-term in the oscillator's energy, independent of temperature. The appearance of this "zero-point energy" led to various speculations.[17] As pointed out by Sommerfeld ([Eckert (2015)], p. 2067), Planck's emission theory was not a progress in itself, but showed that his radiation formula could be obtained from different hypotheses. As for Planck's reformulation of quantum statistics, it was named by Heilbron and Kuhn as one of the sources of the phase-integral condition, independently invented in 1915 by Sommerfeld, William Wilson and Jun Ishiwara ([Heilbron and Kuhn (1969)], note 156 on p. 280), In fact, despite its poor success, Planck's second theory contained ingredients that would reappear fifteen years later with the formulation of quantum mechanics: the zero-point energy of a quantized oscillator and the presence of elementary processes subject to a probability law.

For reasons that will become clear in later sections, we quote Planck on his quantum of action h, and on Einstein's controversial light quantum hypothesis ([Solvay I (1912)], p. 100):

> Let us examine more closely the question of the physical nature of the constant h. First of all a question of principle arises: does this element of action have a physical significance for the propagation of radiant energy in vacuum, or does it only intervene by its very nature in the phenomena of production and destruction of radiant energy in emission and absorption? Depending on the answer given to this question, the further development of the theory will have to follow completely different paths. The first point of view was adopted by A. Einstein with

[16]Lorentz had drawn attention to such contradictions in his *Wolfskehl lectures* dated October 1910.

[17]The idea of a "zero point energy" was at the origin of Einstein and Otto Stern's ill-fated paper of January 1913. We will return to this matter when discussing some highlights of Solvay II.

> his hypothesis of light quanta, and he was followed by J. Stark. According to this hypothesis, the energy of a light ray of frequency ν is not distributed continuously in space, but propagates in a straight line by determined quanta of magnitude $h\nu$, in the same way as the luminous particles in Newton's theory. One invokes, as a confirmation of this hypothesis, the fact that the velocity of the secondary cathode rays, produced by Röntgen rays, is independent of the intensity of these rays. J. J. Thomson found himself led to a similar conception by the study of photoelectric phenomena. He believed he could only explain the small number of electrons emitted, and the independence between their velocity and the intensity of the incident light, by admitting, instead of a uniform distribution of energy in the front of light waves, local accumulations of this energy.[18] It goes without saying that such hypotheses are irreconcilable with James Clerk Maxwell's equations and with all the electromagnetic theories of light proposed so far. When we think of the complete experimental confirmation that Maxwell's electrodynamics received through the most delicate interference phenomena, when we think of the extraordinary difficulties that its abandonment would entail for the entire theory of electrical and magnetic phenomena, we feel some reluctance to ruin its foundation at first glance. For this reason we will leave aside, in what follows, the hypothesis of light quanta, especially since its development still remains quite primitive.

We will see at the end of this overview what Einstein had to say on the difficulty of admitting the reality of light quanta. As for J. J. Thomson's idea of a "non-uniform distribution of the energy in light waves", we have reasons to believe that it inspired Maurice de Broglie in 1922, when he tried to interpret the results of his experiments (see Sect. II.3.7). Right now, it is instructive to notice that Planck made a distinction between the "optical" photoelectric effect, a puzzling phenomenon observed in the case of light, and the ability of Röntgen rays to produce "secondary" electrons.[19] As we will see, it would only take a few more months, and Max Laue's discovery of X-ray diffraction, for physicists to recognise that X-rays are nothing else than a high frequency form of light.

---

[18]Planck quoted J. J. Thomson's Adamson Lecture at Manchester University *Electricity and Ether* [J.J. Thomson (1908)].

[19]We will see that Sommerfeld regarded X-rays as electromagnetic pulses caused by the stopping of cathode rays.

I.1.4.2 *The quantum of action and atomic dimensions*

During the discussion of Planck's report, attention was paid to the work of Arthur E. Haas, the first physicist to propose a connection between atomic dimensions and the quantum of action. Planck expressed criticism against the Haas model, arguing that it was based on classical dynamics, and could therefore only lead to the unacceptable Rayleigh–Jeans radiation law. Lorentz rejected the argument on the grounds that Haas had resorted to non-classical actions. As for Sommerfeld, he entered the discussion, declaring that a "general" quantum hypothesis was to be preferred to any particular atomic model. Lorentz closed the debate with the following observation ([Solvay I (1912)], p. 124):

> Mr. Sommerfeld does not deny that there is a relation between the magnitude of the constant h and atomic dimensions (such as those of the positive Thomson spheres). One can express this in two ways: the constant h is determined by these atomic dimensions (Haas), or the dimensions one attributes to atoms depend on the magnitude of h. I do not see any great difference.

I.1.4.3 *Sommerfeld's h-hypothesis*

Another theoretical novelty was the *h-hypothesis* ([Eckert (2015)], p. 2067), presented by Sommerfeld as an alternative to Planck's quantum hypothesis. Sommerfeld's quest for a general theory, based on the quantum of action, was motivated by the need of a new theory, applicable to individual non-periodic molecular processes, and his belief that these processes are governed by a quantity with the dimension of an *action* (the product of the energy exchanged in the process by the time necessary for the exchange). Sommerfeld set out to test his *h-hypothesis* on a series of non-periodic phenomena (the production of polarised X-rays, the radioactive emission of γ rays, the photoelectric effect, and the production by X-rays of secondary cathode rays). Regarding the first phenomenon, he considered the ratio of the energy of polarised X-rays to the energy of the cathode rays, and showed that his hypothesis, applied to the duration of the stopping of the cathode particles, allowed him to obtain a value of this ratio in fairly good agreement with the experimental value.[20] Turning to γ-rays, Sommerfeld

[20] Sommerfeld used an experimental value obtained by means of a method imagined by Wien and published in [Wien (1905)].

noted that their electromagnetic nature had recently been confirmed by a study of the fluctuations in the ionisation current produced by these rays. However, he immediately added[21] ([Solvay I (1912)], p. 334):

> Egon von Schweidler, to whom the discovery of radioactive fluctuations is due, pointed out that measuring these fluctuations should make it possible to decide between the corpuscular and the electromagnetic theory of γ-rays. Edgar Meyer, after carrying out such measurements, concluded that, for the distribution of energy, γ-rays behave more or less in agreement with the corpuscular theory.

Then, after an examination of the electromagnetic theory, according to which γ rays accompany the emission of high velocity β-particles, Sommerfeld concluded: "I found that ... for high velocities of the emitting electron, the electromagnetic radiation takes on the character of a Newton luminous projectile, the more so when the velocity of the β-particle gets close to that of light."

I.1.4.4 *On γ and β rays*

Regarding the connection between γ and β-rays, Sommerfeld showed that his hypothesis allowed him to compare two ratios: that of the pulsation energy of a γ-ray to the kinetic energy of the corresponding β-ray, and that of the ionising powers of these rays, measured in the same ionisation chamber. He also pointed at a major problem of ordinary electrodynamics (its difficulty in accounting for the large concentrations of energy that manifest themselves in the photoelectric effect, and even more so in the X-ray emission of secondary cathode rays) and noted that this difficulty seemed to disappear if one assumes that a quantum law, to which the h-hypothesis applies, governs the liberation of an electron from its atomic bond. Focusing on the photoelectric effect, Sommerfeld showed that his theory provided a simple explanation "of the accumulation of a quantum $h\nu$, borrowed from the energy of the continuous incident radiation". He also observed that his concept of the "quantum of action" provided a better account of photoelectric data than Einstein's theory of light quanta.[22]

[21]Sommerfeld is referring to Schweidler's communication presented in Liège on September 1905.

[22]Sommerfeld mentioned that, according to Wright [Wright (1911)], the maximum photoelectric energy did not vary linearly with frequency. He also indicated that the photoelectric effect was selective: it showed an influence of the atom's eigenfrequencies.

Finally, after comparing his hypothesis to Planck's hypotheses, Sommerfeld recognised the difficulty in applying his theory to situations more complex than the above elementary processes. He insisted on the hypothetical character of the results obtained so far, and announced that further tests of his theory were in preparation in his Institute.[23]

I.1.4.5 *Rutherford's silence about an atomic nucleus*

More surprising was Rutherford's prudent attitude during the discussion of Sommerfeld's report. Marie Curie insisted on the distinction that should be made between two regions in the atom: an internal region and a peripheral one. She based the idea on the results of scattering experiments involving α and β particles. Curiously, Rutherford made no mention of the Manchester α-scattering experiments, and the findings of Ernest Marsden and Hans Geiger suggesting the existence of an atomic nucleus.

During his investigations, on which he reported in May 1911 [Rutherford (1911)], Rutherford had been impressed by Geiger's observation that when α particles passed through a mica film, a small number bounced back. He would have said ([Andrade (1964)], p. 111): "It was quite the most incredible event that ever happened to me in my life. It was as incredible as if you fired a 15-inch shell at a piece of tissue paper and it came back and hit you!" Rutherford's silence on his nuclear atom during the discussion of Sommerfeld's report was emphasised by several authors, in particular by Pais ([Pais (1991)], p. 125):

> He (Rutherford) remained silent about his model during a major international conference (the first Solvay conference) held in the fall of 1911. In his 670 pages book, completed a year later, only 3 pages are devoted to α particle scattering.

Several elements can be put forward to explain Rutherford's silence. One is the context in which his model was conceived. Following up on a suggestion from Rutherford, Geiger and Ernest Marsden looked for and observed very large deflections of a narrow beam of α particles passing

[23] It was the need of such tests that prompted Sommerfeld to invite Walter Friedrich, a student of Röntgen, to Munich, an initiative that would be rich in consequences (see [Eckert (2012a)], p. 32).

through a metal foil ([Marsden (1954)], p. 296). Rutherford then suggested to round off the experiments in a form suitable for publication. His reaction to their results was the following ([Marsden (1954)], p. 296):

> He [Rutherford] pondered over this remarkable results for weeks. Such ponderings and related calculations were generally made on the quiet of his study at home but out of it emerged his inspiration of a nuclear atom with positive charge of electricity in a very small nucleus at the centre and electrons in outer planetary orbits somewhat similar in relative spatial dimensions to a solar system in miniature—*a brave conception in view of its reversal of the ideas current at the time in which the positive electricity of atoms was considered as dispersed in a relatively large sphere.* [emphasis added]

As pointed out by Marsden, the assumption of an atomic nucleus was in sharp contradiction with the ideas of J. J. Thomson, the successful designer of atomic models. This explains, at least partially, Rutherford's wish "to consolidate his theory" ([Marsden (1954)], p. 297), before publicly criticising Thomson's model. The additional evidence that Rutherford needed in support of his "nucleus theory" was the observation that it gives the same number of electrons in the atom as the scattering of X-rays (see in section I.5.1 how Rutherford reacted to J. J. Thomson's report at Solvay II).

On the other hand, Rutherford knew that any planetary model would be unstable from the point of view of classical electrodynamics ([Marsden (1954)], p. 297). This awareness is evidenced by the fact that he carefully avoided introducing electron orbits in his paper of May 1911, while remaining vague about the positive character of the charge on the nucleus.

Charles G. Darwin, another member of Rutherford's team, described the situation in Manchester before the first Solvay Council ([Darwin (1962)], p. 5):

> Rutherford recognised that there must be an effect of single scattering ... In 1911 he tried the idea of a heavy central electric charge ... It was, I think, several months before it was called the nucleus ... At this very early stage, even before the theory was verified at all, he could see many further steps it might lead to.

> *However, all this depended on the verification of the main hypothesis, and this verification was undertaken in a laborious series of experiments by Geiger and Marsden.* [emphasis added]

Today we know that decisive evidence in favour of a nuclear atom only arose in 1913 from Moseley's X-ray experiments. Hence, Rutherford's caution at Solvay I may be regarded as proof of his awareness that essential steps forward in science never occur at a single moment. The process leading to the replacement of Thomson's model by Rutherford's nuclear atom started in 1909 with the first experiments in Manchester, and ended towards the end of 1913 ([Darwin (1962)], p. 6).[24]

As a matter of fact, Rutherford exercised restraint during most of the Council debates. He had no appetite for the daring proposals of his German colleagues. True to his distrust of theoretical speculation, he gave this critical account of the Solvay discussions in a letter to W. H. Bragg of December 20, 1911 ([Eve (1939)], p. 208):

> I was rather struck in Brussels by the fact that continental people do not seem in the least interested to form a physical idea of the basis of Planck's theory. They are quite content to explain everything on a certain assumption, and do not worry their head about the real cause of the thing. I must, I think, say that the English point of view is much more physical and much to be preferred.

I.1.4.6 *Discovery of superconductivity*

A spectacular event took place during the discussion of Nernst's report. Kamerlingh Onnes presented a graph that had been obtained in his laboratory a few days before the Council. It showed the collapse of the electrical resistance of mercury at a temperature of 4K (kelvin).[25] With this *scoop* of November 2, 1911, the Solvay meeting became the forum at which Kamerlingh Onnes presented for the first time the phenomenon of *superconductivity*. Prior to the Council, the collapse of electrical resistance at a temperature above the absolute zero had been the subject of two communications to the Amsterdam Academy of Science, but, as pointed out by Hendrik Casimir, there had been a significant evolution in Kamerlingh

[24] We will have more to say about Moseley's experiments in the following chapters.

[25] We remember that according to the International System of Units, the temperature is measured using Lord Kelvin's absolute scale and the temperature unit is the kelvin.

Onnes's understanding of the phenomenon since its first observation on April 8, 1911 ([Casimir (1955)], p. 118):

> It seems that Kamerlingh Onnes did not at once realise the revolutionary character of his observation. He had previously given a formula, according to which the resistance of all pure metals should continuously drop toward zero with decreasing temperature, and was inclined to regard the results for mercury as a confirmation of this very sketchy theory. But at the Solvay Conference in November of that year, and in a communication to the Royal Academy at Amsterdam on November 25th, the abruptness of the transition is emphasised and the idea of superconductivity as a particular condition of matter is from then on established.

Heike Kamerlingh Onnes (1853–1926)
Courtesy of the Solvay Institutes

As for Lorentz, he immediately grasped the quantum significance of the phenomenon observed by his Leiden colleague. Having accepted to chair the

Council in May 1911, he immediately insisted on adding Kamerlingh Onnes to the list of invitees. The name of this Dutch physicist did not appear on the list that had been drawn up on March 13, 1911. On May 31, Nernst mentioned Lorentz's wish to invite his colleague. On June 23, Kamerlingh Onnes was officially cited as one of the invitees to Solvay I.[26] However, he was not invited to present a report. Despite this fact, Lorentz decided to include his communication in the Council Minutes, and to present it as a genuine Solvay report.[27]

I.1.4.7 *Wilson's chamber – Planck's afterthought*

In his report on the reality of atoms, Perrin cited the experiments carried out by C. T. R. Wilson on an ionised gas, free of dust, but saturated with water vapour. The results proved that these ions serve as condensation centres for the droplets of the cloud which forms when a gas is cooled by sufficient adiabatic expansion. Wilson's cloud chamber, a device developed in 1911, was destined, as we will see, to play a major role in the quantum revolution.

Another noteworthy event took place during the discussion of Nernst's report. Planck, as we know ([Klein (1965)], p. 200), sought to establish a link between the quantum of action and the heat theorem. In an attempt to support Nernst's claim about the quantum basis of his theorem, he made this striking statement: "It seems that an explanation of Mr. Nernst's theorem can be found in the theory of quanta." The statement appears in the Notes taken on the spot by Maurice de Broglie. It is still present in the draft report, submitted to the contributors on December 23, 1911, but does not appear in the official volume published in 1912.[28] Its removal could be explained by the fact that Planck's statement was disputed by Einstein, and that Lorentz followed suit after hearing Einstein's objections. We know that in December 1911, the editors of the volume (Paul Langevin and Maurice de Broglie) told the contributors that they could mention remarks they had not made during the meeting, which would then be recorded in the form of a footnote. We also know that Poincaré took advantage of this provision

[26]See the following archival documents: Archive des Instituts Solvay, S.a.b. ULB, doc. 1692, doc. 1696, and doc. 1702.

[27]This slight distortion of the truth explains the absence in the official volume of any real discussion of the "Kamerlingh Onnes report". Only one question was asked (by Langevin), giving rise to only one comment.

[28]See the "Solvay file", in the Archives of the Collège de France, Fonds Brillouin. One of the authors thanks Mrs. A. Chatellier for giving him access to this file.

and corrected the position he defended in Brussels.[29] Hence, we assume that Planck felt authorised to request the deletion of his disputed claim.

I.1.4.8 *Einstein's report*

Einstein divided his report into four parts. In part I, he discussed the relationship between specific heats and Planck's radiation formula. Having recalled that measurements had long shown the inability of molecular mechanics to account for the behaviour of specific heats at low temperatures, he made the following comment:

> Until a few years ago, we had no possibility of relating this insufficiency to other physical properties of matter. Planck's research on thermal radiation shed unexpected light on this point. Although we are not able to replace classical mechanics with another mechanics, which gives correct results, even for rapid thermal movements, we can at least predict the law according to which deviations occur from the law of Dulong and Petit, and quantitatively relate these deviations to other physical properties of the substance.

With regard to the "Nernst–Lindemann" formula, presented by Nernst as an improvement on Einstein's formula, Einstein recognised its ability to account for the measured values of the specific heats of solids.[30] However, he rejected the idea that the formula could be given a theoretical meaning.

But that was not all. Far from conveying the positive message that Nernst was hoping for, Einstein did not insist on the success of Planck's hypothesis in the molecular field. He indicated that Planck's conceptions could sometimes lead to results that were contradicted by experiments, and concluded:

> Hence, it is necessary to accept only with distrust any new application of the method which consists of deducing the thermal properties of matter from the radiation formula, since in any

[29]The conclusion of Poincaré's in depth-analysis of the status of Planck's hypothesis, made shortly after the meeting, can be found in the Gauthier–Villars volume. See ([Solvay I (1912)], footnote on p. 453).

[30]The Nernst–Lindemann formula had been obtained by assuming that the substance could be treated as a mixture of Planck's oscillators, half of which oscillated with a frequency ν, the other with a frequency $\frac{\nu}{2}$.

> application of this kind one must rely on mechanics, the validity of which is certainly not general, and on electrodynamics which will probably also have to be abandoned.

In Part II, Einstein addressed the crucial, but unresolved question: "How must we modify Mechanics to bring it into agreement with the radiation law and with the thermal properties of matter?" After recalling Planck's work on the radiation law, he showed that in its current form quantum theory could not explain the behaviour of thermal conductivity of matter in the vicinity of the absolute zero of temperature, i.e. zero kelvin. Faced with this difficulty, Einstein sought to draw from the thermal properties of matter conclusions on the statistical laws of thermal phenomena. Applying Ludwig Boltzmann's theorem to a solid body in thermal contact with a reservoir with infinite heat capacity, he calculated the mean square of the deviations of the body's energy from its average value with the help of Planck's radiation law. He thus obtained a *fluctuation formula* consisting of *two* terms.[31] One term corresponded to the predictions of the wave theory of radiation. The other could be explained by assuming radiant energy "to consist of localised quanta". In his report, Einstein declared: "This way of seeing seems completely irreconcilable with the phenomena of diffraction and interference." In short, Einstein's conclusion left no doubt about the depth of the crisis: "In any case, it seems from these considerations that our Electrodynamics cannot, any more than our Mechanics, be brought into agreement with the facts." To drive the point home, Einstein ended part II of his report with the following remark: "I had no other goal here than to show how fundamental are the difficulties in which the radiation formula leads us, even if we consider it as a simple experimental fact."

In Part III, devoted to the quantum hypothesis and its application to more general situations, Einstein emphasised a point he considered well established:

> When a body exchanges energy by a quasi-periodic mechanism of frequency ν, the statistical properties of the phenomenon are the same as if the energy moved by integer quanta of magnitude $h\nu$. Although we do not see by what mechanism this property can be explained, we must admit that the disappearance of periodic energy occurs by quanta of magnitude $h\nu$, and also that the energy must

[31] Einstein had already derived this formula in 1909. See ([Einstein (1909)], p. 189).

be available by similar quanta for periodic phenomena of frequency ν to occur.

Turning his attention to specific processes, such as photochemical reactions and the photoelectric effect, Einstein noted that the facts in no way corresponded to the predictions of ordinary theory:

> One would expect that a certain minimum density of electromagnetic energy is necessary to cause the photochemical rupture of a molecule. On the other hand, one does not see why radiations of high frequency can produce phenomena of greater energy than lower frequency radiations. We do not understand the specific action of frequency any more than the absence of action related to intensity. In particular in the photoelectric effect, the order of magnitude of the kinetic energy of the emitted electrons agrees with a quantity "$h\nu$", specific to the incident rays. Faced with these facts, and in particular with the large fluctuations in the conductivity of the air traversed by γ-rays,[32] we can hardly escape the conclusion that, during the absorption of radiation, the energy is transferred by quanta of considerable magnitude, and that the borrowing of energy by secondary rays cannot occur continuously in space and time. These discontinuities which make it so difficult to accept Planck's theory really seem to exist in nature. The difficulties that a satisfactory theory of these fundamental phenomena must overcome seem, at this moment, insurmountable. Why does an electron in a metal struck by Röntgen rays, take the high kinetic energy observed in the secondary cathode rays? All the metal is in the field of the Röntgen rays; why is it only a small part of the electrons that pick up the speed of these cathode rays? How come that the energy is absorbed in extraordinary few points? How do these few points differ from others? We have no answer to these questions and to so many others.

In the last part of his report, Einstein considered the attempts, made by Nernst and by Sommerfeld to link Planck's constant h to the mechanical properties of atoms or electrons. His goal was to see whether their results could be deduced from the radiation formula. He first considered Nernst's

[32] Regarding this aspect, see section I.6.3 in which we discuss the Solvay subsidy granted to Edgar Meyer.

result for the variation with temperature of the rotational energy of the molecules of a diatomic gas. Applying Planck's expression for the energy of a linear oscillator to a model in which all molecules are supposed to rotate with the same frequency, he obtained a result similar to that of Nernst. Einstein then focused on Sommerfeld's hypothesis on elementary shocks. Considering a mono-atomic gas in thermal equilibrium with a black-body radiation field, he came to the following conclusion: "In order to be able to deduce Sommerfeld's hypothesis from the radiation formula it suffices to admit that the electron theory is correct with regard to the emission of radiation."

I.1.5 Conclusions and long-term effects of Solvay I

Various statements reflect the general feeling during and at the end of the Council. Here is a statement on "Quantum Theory", made by Poincaré in chapter VI of his Last Essays ([Poincaré (1963)], p. 75)[33]:

> One may wonder if mechanics is not on the eve of a new commotion. A congress of about twenty physicists from different countries assembled recently in Brussels, and at every moment they could have been heard talking of the new mechanics which they contrasted with the old mechanics. Now, what was that old mechanics? Was it that of Newton, the one which still reigned uncontested at the close of the nineteenth century? No, it was the mechanics of Lorentz, the one dealing with the principle of relativity; the one which, hardly five years ago, seemed to be the height of boldness.

Poincaré died on July 17, 1912. He was therefore prevented from witnessing the development of a theory he had put on a solid base. This point was underlined by McCormmach ([McCormmach (1967)], p. 51): "There can be no question that Poincaré and quanta offer an instance of one of the more dramatic confrontations and responses in the history of science."

Another statement on the situation towards the end of Solvay I was made by Einstein during the discussion of his report:

[33]Poincaré died in 1912. His last letters were published in 1913, and republished fifty years later in 1963.

> We all agree that in its current form quantum theory can be useful, but that it does not constitute a theory in the ordinary sense of the term, or at least not a theory that from now on can be developed in a coherent manner. On the other hand, it is well established that classical dynamics, translated by the Lagrange–Hamilton equations, cannot be considered as providing a sufficient scheme for the theoretical representation of all physical phenomena.

Marcel Brillouin (1854–1948)
Courtesy of the Solvay Institutes

A final and striking statement was made by Marcel Brillouin during the final discussion:

> It now seems quite certain that it will be necessary to introduce into our physical and chemical concepts a discontinuity, an element varying by leaps, of which we had no idea a few years ago. How should it be introduced? This is what I do not see. Will it be in the form first proposed by Mr. Planck, despite the difficulties it raises, or in the second form? Will it be in Mr. Sommerfeld's form, or in some other form still to be found? I don't know ... The uncertainty in which we find ourselves about the form and extent of

> the transformation that must be operated – evolution or complete overhaul – is a powerful stimulus. This concern will certainly haunt us for many weeks, and each of us will passionately focus on the solution of the difficulties that our discussions have shown to be inevitable and important in so many fields of physics and chemistry ... I believe that it is really a new element which imposes itself on our reflexions. Something that has hitherto been treated as varying with continuity now appears to vary in certain circumstances only by sudden increments. Which quantity is it? Under what circumstances does variation cease to be continuous? We haven't figured it out yet. I should add that it is not very satisfactory to be reduced to recognising the presence of a discontinuity by means of apparently continuous phenomena, to introduce it at the basis of a theory, before drowning it out with the help of statistical considerations... If we could imagine an experiment which made us seize the discontinuity "on the spot," it would be much more decisive and instructive.

As we will see (section II.3.5), it would take physicists almost ten years to respond to Brillouin's wish.

Most of the Council members seemed ready to align with the Frenchman's position. As noted by Armin Hermann ([Hermann (1971)], p. 139), only two Council members, Jeans and Poincaré, "took a clearly negative position with respect to the quantum concept ... Poincaré, on the other hand changed his opinion as a result of the conference."

As for two representatives of the next generation – Niels Bohr and Louis de Broglie – they had the privilege of obtaining first hand information on the debates that had taken place at Solvay I.

In 1961, Bohr recalled his meeting with Rutherford in November 1911 ([Solvay XII (1962)], p. 17):

> A vivid account of the discussions at the first Solvay meeting I got from Rutherford, when I met him in Manchester in 1911, shortly after his return from Brussels. On that occasion, however, Rutherford did not tell me, what I only realised some months ago by looking through the report of the meeting, that no mention was made during the discussions at the conference of a recent event which was to influence the following development so deeply, namely his own discovery of the atomic nucleus.

Louis de Broglie, on his part, accompanied his brother to Brussels, in order to assist him in his duties as Scientific Secretary of the Council.[34] Having become a physicist himself, it did not take him long to make his mark on the quantum revolution. Indeed, in November 1924 Louis defended a thesis in which Einstein's controversial views on the dual nature of light were boldly extended to the realm of matter. Thirty years later, he recalled that it was after reading the reports and discussions of Solvay I that he decided to become a physicist, with the firm intention of clarifying the mystery of Planck's quanta ([L. de Broglie (1953)], p. 458):

> My brother who had been one of the secretaries of the Council, and was preparing the publication of its proceedings, showed me the texts of the discussions. With the ardour of my age, I got filled with enthusiasm for the problems under study, and I promised myself to concentrate all my efforts on understanding the true nature of the mysterious quanta that Max Planck had introduced into physics ten years earlier, but whose deep meaning was not yet perceived.[35] From that time, I realised the importance that classical formulations of analytical mechanics, and in particular those of Hamilton and Jacobi, represented for the study of this question, and I had begun to deepen my knowledge of these theories.

I.1.6 An *international* "Select Committee"

On the last day of the conference, Einstein's report was interrupted to allow the holding of a private meeting of Solvay with the members of a "Select Committee".[36] This Committee, chaired by Lorentz (the Netherlands) comprised Marie Curie and Marcel Brillouin (France), Nernst, and Warburg (Germany), Rutherford (United Kingdom), Kamerlingh Onnes (the Netherlands) and Knudsen (Denmark).

---

[34]Private communication from Philippe-Maurice de Broglie, grandson of Maurice de Broglie. The authors wish to thank the Duke for providing access to the family archives and to Maurice's laboratory notes in Broglie (FR).

[35]Both de Broglie brothers noticed the tension between the observed quantum phenomena and analytical mechanics. As for Maurice, see ([M. de Broglie (1951)], p. 12).

[36]This event, reported in Maurice de Broglie's "Minutes" of the Council, does not appear in the official volume, published by Gauthier-Villars in 1912.

We do not know what was said during the meeting, but there is little doubt that Solvay announced his wish to support new experiments on radioactivity and Brownian motion,[37] and that he considered creating an Institute for this purpose.[38]

It is also likely that Solvay was strengthened in his idea by a statement made by Knudsen at the end of his report on the properties of ideal gases.[39]

Martin Hans Christian Knudsen (1871–1949)
Secretary of the ISC from 1912 to 1930
Courtesy of the Solvay Institutes

[37]Solvay made this clear in his closing speech (see [Lambert and Berends (2021)], p. 77). He needed new evidence in support of his views on the seemingly inexhaustible amount of energy emitted by radium (he thought that radium had the capacity to extract energy from the environment, an idea initially shared by Pierre Curie and Lord Kelvin).

[38]We will see that the news of a Solvay Physics Institute spread from November 1911, and reached Moscow before March 1912.

[39]We know from various sources that Solvay was briefed every evening by Lorentz on the events that had taken place at the Council.

These were his words translated into English ([Solvay I (1912)], p. 455):

> The data concerning the thermal conductivity of gases are much rarer and even more discordant, and it seems necessary to me to complete and improve them by a systematic examination of the various gases, for various temperature and under various pressures, and by a comparison of the various methods. This task seems to me too important and too difficult to be carried out in different laboratories, using different methods, and it would perhaps be desirable to entrust this research to one of the institutes where it could be carried out, or to create a new institute for the solution of this and other questions dealt with by our Scientific Council.

Before ending this chapter, it seems appropriate to devote a few words to the group photo that was taken at Hotel Metropole. The vague appearance of Planck's radiation law on the blackboard may be regarded as an indication that the picture was taken on October 31, 1911, shortly after Planck's presentation. Unlike Rutherford, who stares at the photographer, Marie Curie and Poincaré seem to be in deep discussion. We do not know what they were talking about, but it is probable that they discussed some particular points of Planck's report. Indeed, they both intervened significantly during its discussion, a fact that corroborates the idea that the picture was taken after the discussion of Planck's report. However, it is also possible that the "art photographer" Benjamin Couprie asked the members sitting next to the table to pretend to be working. This is suggested by Perrin, who seems immersed in some reading. As for Poincaré and Marie Curie, they could have complied with the photographer's request, while continuing a discussion that had begun before the photo was taken. The fact that Solvay appears at the end of the table is proof that the photo was retouched at a later stage: we know that Solvay did not take part in the Council sessions.

Chapter I.2

The International Physics Institute

Hendrick Antoon Lorentz (1853–1928)
Courtesy of the Solvay Institutes

I.2.1 Ostwald's proposal for chemistry

The idea of an "international scientific foundation" had been suggested to Solvay several months before the Council. In the beginning of April 1911, he had been approached by Wilhelm Ostwald, the newly elected president of the International Association of Chemical Societies, who requested his help for the creation of an International Chemistry Institute (ICI).

Genuinely interested in the project, Solvay had promised to devote 250,000 francs to its realisation, in case of agreement on the foundation conditions.[1] Unfortunately, the request came at a bad time, especially since Ostwald wanted Solvay to take an active part in the project. Being totally taken by the prospect of the Physics Council, and the need to complete his *Gravito-Materialitic* theory, the industrialist asked his collaborators to contact Ostwald and inquire about his plans.

By a curious coincidence, it was during the Council that Solvay received a *Memorandum* from Ostwald with details on the Chemistry Institute. Having no choice but to wait and see what he would do with the document, he put it in a drawer. It was also during the Council that Goldschmidt came up with the idea of an *International Solvay Institute of Physics*. The proposal seemed sensible, but slightly premature. Solvay did not feel ready to talk about it in his closing speech. Addressing his guests on November 3, he invited them to meet again in 1913, and added:

> In the mean time, if I could express a wish, it would be to see a follow-up to the experiments that I have in mind on the search for the origin of Brownian and radioactivity energy, in the firm thought that I persist in having that this energy comes from the external environment, and not from the Brownian environment or from the radioactive bodies themselves.[2]

I.2.2 Lorentz and the Solvay Science Project

Solvay's first opportunity to discuss his foundation plans with Lorentz was at a lunch at his house on November 4, 1911. It is likely that the industrialist recalled his wish to have new experiments "carried out under his control". From later correspondence, we know that Lorentz welcomed

---

[1] One of the conditions set by Solvay was that ICI's headquarters be located in Brussels.

[2] Sticking to this old view, Solvay did not accept the growing consensus that radioactivity had its origin in the inner part of certain atoms.

the idea of an experimental programme set up by Solvay, but that he insisted that the programme be extended to other burning issues, such as those raised by the Council.

There are also clear indications[3] that Solvay spoke about his foundation project for physics, and that he specified the missions he intended to entrust to the new Institute. At the international level, its action would be twofold: on the one hand it would grant research subsidies to physicists "from everywhere", on the other it would organise Physics Councils, similar to the one that had just taken place. Solvay also specified the Institute's operating mode. In order to guarantee the "objectivity" of the subsidy programme, and to make sure that the aid be granted to the most qualified applicants, he thought of entrusting the selection process to an *International Scientific Committee*. This Committee would also be responsible for choosing the subjects of future Councils, and establishing the lists of invitees. Regarding its action at the local level, the Institute would grant scholarships to young Belgian researchers wishing to perfect their skills in foreign laboratories.

We do not know if during the lunch of November 4, Solvay mentioned the existence of the Ostwald proposal for an International Chemistry Institute. However, it is clear that it was either in Brussels, or a few days later in Leiden[4], that Solvay asked Lorentz to prepare a project for the Physics Institute, and that he gave him Ostwald's "Memorandum" as a possible source of inspiration.

On November 30, Lorentz sent this message to Wien ([Kox (2009)], doc. 230):

> I hope you have recovered from your efforts in Brussels. I have almost regretted taking the floor so often ... After returning home a week ago, I had to edit all the remarks, so I now think back to our conference with unmixed pleasure. I feel I gained a lot from it, and I only regret that, because of the presidency, I could not enjoy more of the personal contacts with you and other professionals. I found a lot of work here, so I have hardly gotten around to working out a plan for the foundation wanted by Mr. Solvay. I hope, however, to lay my hands on it now and will try to do so, to deal with this matter at a pace that is not too slow.

[3]See Lorentz's correspondence with Solvay of January 1912, Archives of the Solvay Institutes, S.a.b.ULB.

[4]Solvay had been invited to visit Kamerlingh Onnes's cryogenic laboratory.

Wien replied two days later ([Kox (2009)], doc. 231):

> I also think back to the days in Brussels with great pleasure. I received great inspiration there and I am now giving a private seminar to my students on the issues discussed in Brussels. Fortunately, I had already finished editing the discussions here. The new Solvay Foundation can certainly be of great benefit. It is not excluded that I myself will approach this foundation for my experimental work. The demands on the apparatus are now growing so fast that it is difficult even for well equipped laboratories to keep up.

We conclude from the above exchange that by the beginning of December 1911, Wien, a Council member who did not belong to the *Select Committee*, was fully aware of Solvay's plan to create an Institute that would provide subsidies to researchers working in foreign countries.

I.2.3 The International Scientific Committee

Having accepted the mission entrusted to him, Lorentz communicated his conclusions in a letter to Solvay, dated January 4, 1912.[5] After pointing out that the Ostwald proposal did not meet the founding conditions for a Physics Institute, he presented a plan for an *International Solvay Institute of Physics* (ISIP), with a "two-headed" management.[6]

ISIP's scientific direction would be entrusted to the *International Scientific Committee* (ISC) mentioned by Solvay. Its composition would change over time, so as to maintain a close contact with the latest developments in physics. Administrative and financial matters would be the responsibility of a local *Administrative Commission* (AC).

Aware of Solvay's desire to speed up ISIP's subsidy programme, Lorentz considered giving a role to the *Select Committee* he had set up during the Council at Solvay's request.[7] This was his message to Solvay:

> If you agree to entrust the task of the first Scientific Committee to the one that already exists, we could start working as soon as

[5] Letter from Lorentz to Solvay of January 4, 1912, Archives of the Solvay Institutes, S.a.b.ULB, doc.1.

[6] The idea of a two-headed directorate of the Institute had been suggested by Oswald for the ICI. It would be one of the keys to the success of Solvay's international foundations.

[7] Most members of this *Committee* were first-rate experimentalists, a fact which seems to indicate that Solvay intended to call on their expertise for the conduct of the experiments he had in mind.

> you would have taken your decision. I am telling you this because I remember that you expressed yourself on this matter during our conversations in Brussels.

Solvay reacted with enthusiasm in a letter[8] to Lorentz of January 9, 1912:

> As you understood, and as you say in your letter, I'm not only aiming at the progress of science in general; my ambition is to give impetus to the scientific movement in my country. I already worked hard in this direction, either by founding institutes, or by other interventions. And now it is a new effort that I still want to make. I thank you for wanting to help me with your advice: no one is more authorised in this respect than you. Your project pleases me because, in its outline, it responds perfectly to what my life's ambition has been.

I.2.4 Foundation of the International Physics Institute

ISIP's statutes were drawn up in Leiden by Lorentz and by Paul Héger, Solvay's trusted man who headed his Brussels "Physiology Institute". The Physics Institute was founded on May 1, 1912, for a period of 30 years. A capital of one million francs, deposited by Solvay in an Insurance Company, was meant to produce an annual income of 54,000 francs. ISIP's first actions took place in the early summer of 1912. This was made possible by the combination of three factors:

(i) Solvay's acceptance of ISIP's statutes, and his decision of having "nothing to see, nor to say" in the Institute's management.
(ii) The agreement of the members of the Select Committee to take a seat in ISIP's first Scientific Committee.
(iii) Lorentz's availability,[9] and his acceptance to chair the ISC.

The Scientific Committee now comprised the following members: Lorentz (chair), Marie Curie and Brillouin (France), Nernst and Warburg (Germany), Rutherford (United Kingdom), Kamerlingh Onnes (the Netherlands), Knudsen (Denmark), Goldschmidt[10] (Belgium).

[8] Archives of the Solvay Institutes, S.a.b.ULB, doc.2a.

[9] Thanks to unexpected circumstances, Lorentz renounced his ordinary professorship in Leiden and found himself in a position to spend much of his time to ISIP's management (see [Lambert and Berends (2021)], section 3.2).

[10] This Belgian representative was added to the other members in August 1912 (see [Lambert and Berends (2021)], section 4.3).

The Institute's neutrality was exemplary. The chairman of the ISC resided in the Netherlands, a neutral country; the secretary (Knudsen) resided in Copenhagen, capital of a neutral country; ISIP's Administrative Commission was located in Brussels, capital of a third neutral country.

On June 2, 1912, Knudsen sent a letter to the members of the ISC, informing them of ISIP'S creation.[11] The letter specified that one third of the annual budget, about 17,000 francs, would be spent in grants. Lorentz suggested limiting ISIP's support to projects connected to topics that had been highlighted during the Council (X-rays, radioactivity, quantum phenomena, questions related to molecular theory). Committee members were invited to make proposals.[12]

Rutherford reacted[13] on July 8, 1912:

> I quite agree that it is desirable to restrict as far as possible the investigations to which subsidies are given. I certainly think that for the first year the money should be spent to assist investigations on radiation generally. I have no definite proposition to make at the moment but may possibly later raise the question of promoting investigations on characteristic Röntgen radiations and on the emission of β and γ rays from radioactive substances

Rutherford's idea of promoting research on Röntgen rays may come as a surprise. Indeed, as noted by Wheaton with regard to a later episode ([Wheaton (1983)], p. 213), no research on X-rays had been carried out in Manchester before October 1912. Wheaton expands on Moseley's work in Manchester, and his strong desire to investigate X-rays instead of β-rays. In fact, we have reasons to believe that a series of factors – Lorentz's suggestion to support projects connected to the Council topics, the change in Rutherford's interests, Moseley's attitude – were all motivated by the publication in June 1912 of an unexpected event: Max (von) Laue's discovery of X-ray diffraction in crystals (a subject that will occupy us soon). As for Rutherford's specific interest in "characteristic Röntgen rays", it may be explained by his relationship with Charles G. Barkla, the

[11] The letter had been written by Lorentz (Paris, Archives de l'ESPCI, Fonds Langevin, L10/71).

[12] See the Archives of the Solvay Institutes, S.a.b.ULB, docs. 254 and 261.

[13] Letter from Rutherford to Lorentz, Archives de l'ESPCI, Fonds Langevin, L10/236.

greatest British X-ray expert, and celebrated discoverer of the secondary "characteristic" X-rays.[14]

I.2.5 To the rescue of a Russian laboratory

Rumours about Solvay grants started circulating, as we saw, from the end of November 1911. News quickly spread that physicists, engaged in experimental research on "issues of fundamental importance", could apply for support to ISIP. Curiously, the first subsidy request came from Russia. It was made well before the announcement of ISIP's foundation. On February 9, 1912, Lorentz received a message from Piotr Lebedew, a well-known Russian physicists, who alerted him to the dramatic situation of his Moscow laboratory.[15]

The reason was the so-called "Kasso affair" (after the name of Lev A. Kasso, the Minister of Education at the time). Nearly a hundred professors and lecturers at Moscow University had decided to resign after a brutal intervention of the police.[16] Lebedew, who was in this situation, had been forced to appeal to the generosity of Muscovite businessmen. Thanks to their help he had set up his team in private premises. Having been informed of Lorentz's role in a newly created "International Institute", Lebedew had decided to send him a request for support.

Lorentz was extremely surprised. News about grants had clearly leaked from the *Select Committee* or from an ordinary Council member. Someone had communicated Solvay's plans to Lebedew. Lorentz suspected Sommerfeld and sent him a letter which has been lost. However, in his answer ([Kox (2009)], doc.353), Sommerfeld confirmed that he mentioned ISIP to Paul Ehrenfest, an Austrian physicist from St. Petersburg who, after visiting Munich, was due to return to Russia in March 1912. Ehrenfest ([Kox (2018)], doc. 289) acknowledged in a later letter to Lorentz that he had spoken about the Institute to a student of Lebedew.

---

[14] Barkla, who would be awarded the Nobel Prize in 1917, was one of the first beneficiaries of ISIP's subsidy programme in 1912 (see section I.3.2).

[15] Letter from Lebedew to Lorentz, Archives of the ESPCI, Fonds Langevin, doc. L10/232. Lebedew's laboratory was known for the experimental verification of the pressure exerted by radiation.

[16] The intervention was caused by a student meeting on the occasion of the death of Lev Tolstoy. In reaction to the decision of some professors to abandon their seat on the University Council, they had been suspended from their teaching. Most of their colleagues had decided to join them in solidarity.

The mystery being clarified, Lorentz had been tempted to respond immediately to the call of his unfortunate colleague. Unable to do so, he sent a letter to Solvay,[17] praising Lebedew's work, insisting on the dramatic situation of the Moscow physicists, and emphasising the benefits that an International Institute could bring to experimental researchers in a country where leading scientists were deprived of any support from local authorities.

In June 1912, after the setting up of ISIP's Scientific Committee, Lorentz felt free to alert Solvay and the ISC members to the fate of the Moscow physicists. Unfortunately, Lebedew had died on March 1, 1912, and the research carried out under his successor, a Dr. Pyotr Petrovich Lazarev, did not meet the criteria for a Solvay grant. But Lorentz would not give up. Aware of the difficulty of making use of ISIP's subsidy budget, he proposed to take 8,000 francs from the Institute's operating budget for the year 1912–1913 and to grant that sum to Lazarev. He knew he could count on Solvay's support.[18]

In his proposal to provide aid to the Lebedew team, Lorentz argued that a limited donation from an *International Institute*, would encourage Muscovite businessmen to step up their support. Solvay agreed, as did the ISC members. On August 20, 1912, a check of 8,000 francs was sent to Dr. Lazarev. ISIP's first action had a symbolic value: it illustrated the founder's desire to encourage experimental research in physics *conducted in a true scientific spirit, whether in an isolated laboratory or in a recognised centre.*

Meanwhile, the setting up of ISIP's Administrative Commission was in progress. On August 22, Solvay announced his wish to be represented by Emile Tassel, a former professor at l'Université Libre de Bruxelles. He also named Professor Paul Héger, as a candidate for becoming the King's representative. As for the University, Solvay felt confident that it would accept to be represented in the Administrative Commission by Professor Emile Verschaffelt, a Belgian physicist who had worked under the direction of Kamerlingh Onnes. His nomination was confirmed on October 12, 1912.

---

[17]Letter from Lorentz to Solvay of March 6, 1912, Archives of the Solvay Institutes, S.a.b.ULB, doc.115a.

[18]Reacting to the news of Lebedew's death, Solvay had expressed sadness and admiration for the work of the great Russian physicist (see his letter to Lorentz of May 20, 1912, Archives of the Solvay Institutes, S.a.b.ULB, doc.38c).

Chapter I.3

Towards a second Physics Council: The first subsidies

Charles Glover Barkla
Courtesy of the Solvay Institutes

Max von Laue
Courtesy Nobel Foundation archive

The first beneficiaries of Solvay subsidies

I.3.1 Launch of Solvay II

In a letter[1] to a member of the ISC, dated August 22, 1912, Lorentz recalled the support given by ISIP to Dr. Lazarev. Expressing great enthusiasm for the discovery of X-ray diffraction, he stressed the need to provide Laue with additional resources.

On September 23, 1912, Knudsen sent the agenda for the first meetings of ISIP's Scientific Committee, scheduled for September 30 and October 1.[2] The main decisions to be taken concerned the subject of the Physics Council of October 1913, announced by Solvay in his closing speech of November 3, 1911, and the follow up to be given to the subsidy requests that had been submitted to Lorentz during the summer.

Regarding Solvay II, the Committee decided to focus on "The Structure of Matter". With hindsight, it seems that Lorentz and his colleagues foresaw the role that atomic physics was going to play in the quantum revolution. In fact, as noted by Heilbron ([Heilbron (1977)], p. 74), their choice was quite natural:

> Planck, as we know, discovered the quantum in 1900. But, with few exceptions, physicists did not realise the importance of his discovery until about 1910, when as Klein observed, the success of Einstein's theory of specific heat showed that Planck's innovation could be made fruitful outside the subject of radiant heat. As physicists began to seek new applications of the quantum they naturally turned to atomic theory, for atoms were understood as emitters of spectral lines, and "*h*" evidently had something to do with the interaction of radiation and matter.

Indeed, in a letter to Rutherford, Marie Curie insisted on addressing the problem of atomic structure: "The meeting next October will consider the most interesting subject we can find at this time, but unfortunately, we really know very little about the constitution of matter, and I wonder if we will be more advanced in a year" ([Heilbron (2015)], p. 2045). The remark may seem odd: Marie Curie was most certainly aware that Rutherford's nuclear model had been in print over a year. But, as noted by Heilbron, "like most physicists, she did not pay attention to it before Bohr showed what it was good for" ([Heilbron (2015)], p. 2045).

---

[1] Archives de l'ESPCI, Fonds Langevin, LO10/89.

[2] Archives of the Solvay Institutes, S.a.b.ULB, doc. 648.

As for Einstein's theory of specific heats, the treatment of the problem that had led to his formula of 1907, and to its discussion at Solvay I, had been significantly improved by Peter Debye, Max Born and Theodor von Karman [Debye (1912); Born and von Karman (1912)]. It therefore seemed appropriate to schedule a report on the molecular structure of solid bodies.

However, as one may easily understand, there was in the autumn of 1912 a subject likely to overshadow all others: Laue's discovery of X-ray diffraction in crystals. This remarkable success,[3] based on experiments carried out in Munich by Friedrich and Paul Knipping, had been communicated by Sommerfeld in June 1912. The famous "Laue spots" had given rise to numerous comments. They provided indisputable proof that X-rays are electromagnetic radiations of the same nature as light. Einstein had been among the first to recognise the far-reaching implications of the phenomenon. Reacting to the photographs that Laue had sent him, he gave free rein to his enthusiasm in a letter to Ludwig Hopf dated June 12, 1912: "It is the most marvellous thing I have ever seen. Bending at the individual molecules, whose arrangement is thereby revealed. The photograph is sharp, which one hardly would expect in view of thermal agitation" ([Klein *et al.* (1994b)], doc. 308).

As reported by Eckert ([Eckert (2012b)]), other physicists reacted in the same manner:

> The discovery was immediately recognised as a sensation. In England, William Henry Bragg and William Lawrence Bragg, father and son, developed an alternative method by which they confirmed the discovery. Within about a year it became clear that the discovery gave birth to two new sciences: X-ray crystallography and X-ray spectroscopy.

In England, the two Bragg's were not alone in their reaction to the news. According to Wheaton ([Wheaton (1983)], p. 213), as soon as Moseley heard of it, he tried to convince Rutherford that "X-rays were a worthwhile experimental object." In view of what we already mentioned, and of the subsidies that will be discussed in the next section, it is clear that it was not hard for Moseley to convince Rutherford of the need to support X-ray research in Manchester. Moseley and Darwin immediately started investigating X-ray diffraction ([Wheaton (1983)], p. 213).

---

[3]Previous attempts to produce X-ray diffraction had failed. See [Wheaton (1983)], pp. 31–40.

They disagreed with Laue's idea that the interference spots were produced by secondary X-rays, characteristic of the crystal. In fact, the true origin of the Laue spots remained unclear until the end of 1912. This uncertainty played, as we will see, a decisive role in the decisions taken by the ISC with regard to the granting of subsidies.

Seeking to complete the agenda of the second Physics Council, Lorentz thought of inviting Woldemar Voigt, an expert in crystallography. Here is an excerpt of the letter he sent him on November 2, 1912 ([Kox (2009)], doc.247):

> You certainly heard about the "Conseil de physique" which met last year in Brussels at the invitation of Mr. Solvay. Since then an International Physics Institute has been founded, thanks to the million francs donated by Solvay. It will ensure from time to time the holding of a conference similar to that of 1911. The next one will take place at the end of October 1913. It will address questions about the structure of matter. On behalf of the Institute's Scientific Committee (Mrs.Curie, Brillouin, Goldschmidt, Kamerlingh Onnes, Knudsen, Nernst, Rutherford, Warburg and myself), I invite you to take part in this conference. Other invitees are Bragg, Grüneisen, Gouy, Laue, Lenard, Röntgen, J. J. Thomson and Weiss.[4] Are also expected to attend: the participants to the first Council, i.e., the members of the above Committee, plus M. de Broglie, Einstein, Hasenöhrl, Jeans, Langevin, F. A. Lindemann, Perrin, Planck, Rubens, and Sommerfeld ... It goes without saying that I would personally be very happy to welcome you at the conference...

In the letter, Lorentz also mentioned the provisional list of reports[5] to be presented at Solvay II:

(1) Structure of the Atom (report by J. J. Thomson or by Rutherford).
(2) Recently discovered phenomena by Mr. Laue (report by M. Laue).
(3) Pyro-and Piezo-electricity (report by W. Voigt).
(4) Molecular structure of solid bodies (report by A. Einstein or by E. Grüneisen).

[4]The new participants Lorentz is referring to are: Eduard Grüneisen, Louis Georges Gouy, and Philip (von) Lenard.

[5]He commented: "We decided not to stray too far from last year's programme."

The letter specified that Voigt would receive an official invitation from ISIP's Administrative Commission, which had been set up at the end of October 1912.

A note had been circulated to inform the scientific community of the Institute's objectives. The announcement of the new Institute was completed in January 1913 by these lines in *Nature* ([Rutherford (1913a)]), in which Rutherford drew attention to the possibility of obtaining subsidies for experimental research:

> A New International Physics Institute
>
> It (the Institute) will offer an opportunity for scientific men of all nations to get together and to exchange views on questions connected with physics and chemistry, and obtain a consensus of opinion as to the best direction in which grants should be given to extend or deepen our knowledge of special subjects.

I.3.2 First Solvay subsidies (X-ray research)

Having agreed on the subjects that would be discussed at Solvay II, the members of the ISC focused on the question of subsidies. Their first concern was to lay down the rules for the evaluation of projects, and for establishing an order of priority. An objective procedure had been worked out by Lorentz. It was adopted on September 30, 1912, with the understanding that it would be applied to projects submitted before September 15, 1913. The Committee members also examined the requests submitted to Lorentz in response to Knudsen's circular letter of June 1912. Two requests, submitted during the summer, concerned X-ray research. One came from Laue, who requested 4,000 marks for a *Study of the diffraction of Röntgen rays by crystals.* The other emanated from Charles Barkla, an expert on "secondary X-rays", who requested £103 for a *Study of fluorescence phenomena.*

Laue had presumably been tipped by Sommerfeld. His project had been reviewed by Lorentz.[6] Barkla, who had been in contact with Rutherford, wished to continue a research, already in progress, that was partially supported by the British Government.[7] The *Minutes*[8] of the Committee

[6] See document L010/86, Archives de l'ESPCI, Fonds Langevin.

[7] See document L010/105, Archives de l'ESPCI, Fonds Langevin.

[8] *Procès-verbaux des Réunions du Comité Scientifique Solvay. Attribution des Subsides,* Archives de l'ESPCI, Fonds Langevin, L010/196.

meetings reveal that Laue's request was accepted during the first session. Not so for Barkla, whose project raised some doubts. However, on October 1, 1912, the ISC proposed to grant 5,000 francs to Laue and 2,500 francs to Barkla. The Administrative Commission agreed, specifying that Laue deserved support *for his research on the diffraction of Röntgen rays by crystals, a phenomenon discovered by him*, and that Barkla deserved a grant *for continuing his research on the secondary rays that are produced when Röntgen rays are scattered by a generic body.*[9]

Both subsidies were paid on October 21, 1912. Laue spent the Solvay money to set up an X-ray facility in Zurich, similar to the one in Munich. He thanked ISIP in a paper on X-ray diffraction written with Franz Tank, which appeared in *Annalen der Physik* in March 1913.[10] Barkla used the subsidy to cover the costs of research on the absorption of X-rays, and on fluorescent X-ray spectra ([Barkla and Collier (1912)], p. 213), which exceeded the British government grant, obtained through the Royal Society. He thanked the Institute for "a grant in aid for his researches", in a report to the London Physical Society [Barkla and Martyn (1913a)], and renewed his thanks in a paper published in the *Philosophical Magazine.* [Barkla and Martyn (1913b)] Barkla used the remaining money to cover the costs of research on ionisation of gases by X- and β-rays, carried out with Albert J. Philpott. Results of this research also appeared in the *Philosophical Magazine* ([Barkla and Philpot (1913)], p. 856).

As we shall see in Part II (section II.4.1), Barkla's X-ray research became a source of inspiration for the American physicist Arthur Holly Compton, who in 1923 discovered an effect, named after him, which established the physical reality of photons. Laue and Barkla, the first beneficiaries of ISIP's subsidy programme, agreed to report on the use made of their grants before November 1913.

I.3.3 Competing interpretations of X-ray diffraction

The Committee's decision to provide simultaneous support to Laue and Barkla should not come as a surprise. We quote Eckert ([Eckert (2012a)], p. 31):

[9]Note of the Administrative Commission, Archives of the Solvay Institutes, S.a.b.ULB, doc. 261.

[10]Laue wrote the paper with one of his collaborators in Zurich. See [Authier (2013)], p. 194.

> By 1908, Charles Glover Barkla had provided new evidence about the nature of X-rays. He arrived at the conclusion that X-rays appear in two varieties. One sort of X-rays was independent of the material from which they emerged and could be polarised; the other was like fluorescent radiation and could not be polarised. The latter was entirely dependent on the irradiated material and designed as "characteristic radiation" ... Laue did not perceive the crystal as a diffraction grating for the primary beam of X-rays, but expected that the primary X-rays would excite the atoms of the crystal to emit characteristic radiation. He imagined that it were the homogeneous X-rays of this characteristic radiation which produced a diffraction pattern, due to the regular spatial arrangement of the centres from which it was emitted ... But how should these fluorescent X-rays, that were emitted by the crystal's atoms without phase relation, produce an interference pattern? Laue must have asked himself this question. If he did not, Sommerfeld surely would have asked the same question, and Laue would not have been able to offer a satisfying answer.

As noted by Wheaton ([Wheaton (1983)], p. 203), there were other weaknesses in Laue's interpretation of X-ray diffraction:

> Laue had difficulty in explaining why an intense interference pattern was cast by diamond, when carbon, its only elemental constituent, had been shown by Barkla to be a weak emitter of fluorescent X-rays. Even more perplexing was the evidence presented by Friedrich and Knipping that the absorption coefficient of some of the interfering rays from zinc was markedly lower than that found for the characteristic X-rays from zinc.

In short, by ascribing the diffraction phenomenon to Barkla's "secondary" X-rays, Laue locked himself up into a misconception. For a full year he persisted in his belief that only fluorescent X-rays could exhibit interference.

Moseley, as we said, disagreed with Laue's interpretation. He told his mother ([Heilbron (1974)], p. 194): "The men who did the work entirely failed to understand what it meant, and give an explanation which is obviously wrong." And Moseley was not alone in his criticism ([Eckert (2012a)], p. 34):

> Many physicists were puzzled by the Munich discovery, and by the initial explanation as an interference of the crystal's

> own characteristic radiation. Peter Debye, who had served as Sommerfeld's assistant until 1911, remarked that "one should generally not trade merit against luck with such things" ... Debye subsequently began to study how the diffraction spots were affected by the vibration of the atoms in the crystal. Laue's view about heat motion of the lattice atoms, as briefly mentioned in the first publication of the discovery, must only have added to the puzzles about its causes...

In view of these problems there is little doubt that the members of the ISC recognised the need to support new research on characteristic X-rays and on their possible role in the Laue phenomenon.

The decision to provide that support without any delay (i.e. before applying the regular procedure) was probably taken in order to facilitate Laue's installation as extraordinary professor in Zurich (his inaugural lecture entitled *The undulatory theory of Röntgen (X-)rays*[11] had been scheduled for December 14, 1912).

As for Barkla, it is likely that Rutherford pointed, during the ISC meeting, at the wealth of information that characteristic X-rays could provide on the laws that govern atomic constitution. Many years later, Bohr recalled the crucial role that Barkla's discovery had played in his work on the Rutherford atom ([Bohr (1961)], p. 34):

> Early in my stay in Manchester in the spring of 1912, I became convinced that the electronic constitution of the Rutherford atom was governed throughout by the quantum of action. A support for this view was found not only in the fact that Planck's relation appeared approximately applicable to the more loosely bound electrons involved in the chemical and optical properties of the elements, but especially in the tracing of similar relationships as regards the most firmly bound electrons in the atom revealed by the characteristic penetrating radiation discovered by Barkla.

It is worth emphasising that the decisions on the requests submitted by Laue and Barkla were taken before the announcement of a new, and more convincing interpretation of Laue's discovery. Indeed, substantial progress in the understanding of X-ray diffraction was made a few months later with W. L. Bragg's interpretation of the Laue phenomenon.

[11]The original title in German reads: *Die Wellentheorie der Röntgenstrahlen.*

William Lawrence Bragg (1890–1971)
Courtesy of the Solvay Institutes

We keep quoting Eckert ([Eckert (2012a)], p. 35):

> When the Braggs learned about the Munich discovery, they too regarded Laue's explanation as wrong. But their investigation of the phenomenon also started with a misapprehension. They were biased by W. H. Bragg's long-held conviction about a particulate nature of X-rays[12] ... However, the Braggs soon became aware that the phenomenon is indeed a diffraction phenomenon of the

[12]Bragg senior defended a hypothesis according to which X-rays were not electromagnetic pulses but streams of neutral particles. He still held this view when he learned of Laue's results. See [Heilbron (1966)], p. 341 and [Wheaton (1983)], p. 87 and 208.

> primary X-rays. While W. H. Bragg hoped for an explanation that would finally account both for the wave and particle aspect of X-ray phenomena, his son began to adopt the idea of a wave interference, even though entirely different from Laue. On 11 November 1912, Bragg junior reported at Cambridge about the Munich experiment. In his talk, published later in the *Proceedings of the Cambridge Philosophical Society*, he interpreted the outcome of the experiment as an interference of X-rays that are reflected on crystal planes, and introduced what became known as Bragg's law ... Thus, by the end of the year, the younger Bragg had dismissed both Laue's initial explanation of the observed spots in terms of the crystal's characteristic radiation, as well as his father's particle interpretation.

Most importantly, the work of W. L. Bragg, and that of his father on X-ray spectroscopy, opened the way to a systematic study of crystalline structure. It transformed Laue's discovery into one of the most fruitful developments of the early twentieth century.[13]

I.3.4 Reports to be discussed at Solvay II

Invitations to the second Physics Council were sent on February 12, 1913. Invitees included the members of Solvay I and the other eight physicists mentioned by Lorentz in his letter to Voigt.

By the end of April, positive answers had been received from W. H. Bragg, Brillouin, Marie Curie, de Broglie, Einstein, Gouy, Grüneisen, Hasenöhrl, Jeans, Kamerlingh Onnes, Knudsen, Langevin, Laue, Lenard, Lindemann, Nernst, Rubens, Rutherford, Sommerfeld, Voigt, Warburg, Weiss and Wien.

Sir J. J. Thomson took more time in answering, but accepted to present a report. Einstein, on the other hand, made clear that he would not report at Solvay II. Lord Rayleigh, Planck and Perrin apologised for not attending the Council. As for Lenard, who first reacted positively, he ended up declining the invitation.

In view of these defections, Lorentz decided to invite two other British scientists. He first consulted Rutherford. Here is Heilbron's

[13] It is worth recalling that the discovery in 1953 of the structure of DNA published by Francis Crick and James Watson was made in W. L. Bragg's Cavendish Laboratory using X-ray diffraction data obtained by Rosalind Franklin at King's College, London.

account of the British participation in Solvay II ([Heilbron (2015)], p. 2048):

> All the British scientists who had participated in the first Solvay Council, including Rayleigh, received an invitation to the second. Rayleigh declined again, giving no reason. Jeans, Rutherford and J. J. Thomson, invited as a new member, accepted. Lorentz thought the delegation still too sparse and asked Rutherford to choose between Charles Glover Barkla, an expert on X-rays, and William Jackson Pope, an expert on crystallography, to prepare a report complementary to Max von Laue's account of his recent discovery of the diffraction of X-rays by crystals. Rutherford chose Pope, who agreed to come if his collaborator, William Barlow, was invited too. Rutherford also proposed his close colleague William Henry Bragg, who could speak about an alternative to the method invented by Laue.

Thus, by August 1913, the Council's programme looked as follows[14]:

1. *The Structure of the Atom*, by Sir J. J. Thomson.[15]
2. *Die Interferenzerscheinungen an Röntgenstrahlen, hervorgerufen durch das Raumgitter der Kristalle*, by von Laue.
3. *The Reflection of X-Rays and the X-Ray Spectrometer*, by (W. H.) Bragg.
4. *The Relation between Crystalline Structure and Chemical Constitution*, by Barlow and Pope.
5. *Quelques considérations sur la structure des cristaux et l'anisotropie des molécules*, by M. Brillouin.
6. *Über die Abhängigkeit der Pyroelektrizität von der Temperatur*, by Voigt.
7. *Molekulartheorie der festen Körper*, by Grüneisen.

Rutherford's idea of asking W. H. Bragg to present a report had come to fruition. However, Bragg junior was not among the invitees. Given ISIP's limited financial means, it seemed difficult to invite two members of the same family. Priority was given to the elder Bragg, a pioneer in

[14]Contrary to Solvay I, the second Physics Council was organised by an International Institute. Its programme comprised lectures in English, German, and French, a normalisation of a peculiarity of Solvay I.

[15]Rutherford preferred to remain in the background, apparently out of respect for Thomson.

X-ray spectrometry. In fact, as argued by Eckert ([Eckert (2012a)], p. 35), Rutherford may have had a personal reason to invite his colleague:

> At first it seemed as if the Braggs were engaged in a harmonic family collaboration. In a joint paper of 1913 they described how a crystal could be used as an X-ray spectrometer. Rather soon, however, they went along separate paths. Bragg senior made X-ray spectrometry his favourite research topic–an investigation which Henry Moseley and Charles Darwin pursued at the same time in Manchester. When Rutherford asked the elder Bragg to delay his publication on this subject so that his young men had a chance to proceed along the same line of research, Bragg gave in but always felt it was not quite reasonable.

A few days before the start of the Council, Lorentz thought of inviting Robert W. Wood, an American expert in optics he had met in September at the Birmingham meeting of the British Association for the Advancement of Science (BAAS).[16] Knowing that Wood would remain for some time in Paris, Lorentz sent a message to the Secretary of the Administrative Commission,[17] arguing that the presence at Solvay II of a well-known American physicist[18] would attenuate the absence of five invitees: Lenard, Perrin, Planck, Röntgen, and van der Waals. The chairman added that Wood would be asked to present a report on resonance radiation, an idea strongly supported by Warburg and Kamerlingh Onnes. Conclusion: On October 22, 1913, an eighth report was added to the Council's agenda:

Resonance radiation and Resonance Spectra, by R. W. Wood

This last-minute initiative is a first example of a specificity of Lorentz's chairmanship of the ISC: his concern to take advantage of every opportunity to invite members likely to enrich the Council discussions with reports on the latest developments in the chosen field. As we will see, Wood's presentation at Solvay II also played another role: it had a direct impact on a decision taken by the ISC on a subsidy request submitted to Lorentz between February and September 1913.

---

[16] Wood had obtained important results on the absorption of alkali vapours. Wood's "beautiful experiments" are cited by Bohr in his *Rutherford Memorial Lecture* ([Bohr (1961)], p. 54).

[17] Letter from Lorentz to Tassel of October 17, 1913, Solvay Archive, S.a.b. ULB, doc. 1821.

[18] Wood was known for having disproved the existence of a new and mysterious radiation: the so-called "N-rays", allegedly discovered by the French physicist René Blondlot.

Chapter I.4

Bohr and atomic structure

Niels Bohr (1885–1962)
Courtesy of the Solvay Institutes

I.4.1 The "miraculous" stability of matter

A major event with regard to the transition from classical to quantum physics took place a few months before Solvay II: the publication in July 1913 of Bohr's first part of his trilogy entitled *On the Constitution of Atoms and Molecules*. [Bohr (1913b, c, d)] At the time Bohr had already been working for several months at Rutherford's laboratory in Manchester.

In his seminal trilogy, Bohr referred to Solvay I, emphasising that discussions established "the inadequacy of classical electrodynamics in describing the behaviour of systems at atomic level" [Bohr (1913b, c, d)]. Indeed, during the discussion of Sommerfeld's report, Marie Curie made a bold remark. Considering a Thomson-like atom, she argued that the β-ray electrons should come from the inner part of the positive sphere, where they would be circulating at a high speed. Poincaré immediately reacted by asking why these accelerated electrons did not radiate electromagnetic energy ([Solvay I (1912)], p. 385). Poincaré's objection could have been seen by Bohr as a strong reason to renounce classical electrodynamics when dealing with an atom in which electrons are revolving around a nucleus.[1] It might have contributed to the introduction of the revolutionary postulates that enabled Bohr to account for hydrogen's emission spectrum, and to produce a theoretical prediction of Johannes Rydberg's constant. In fact, Bohr's postulates included the existence of stationary atomic states (i.e. orbits in which the electron does *not* radiate electromagnetic energy, avoiding the electromagnetic instability of every preceding planetary model), and the possibility for an electron to *jump* from one stationary state to another by absorption or emission of an energy quantum.

Bohr's belief that atomic theory should be based on elements foreign to usual principles, raises an obvious question: *What was the weakness in the classical approach that justified, in Bohr's eyes, such a radical departure from it?* An answer to this question has been given by Heisenberg in chapter III of [Heisenberg (1969)]. Recalling his first meeting with Bohr, Heisenberg makes the following observation:

> Bohr's starting point was the realisation that classical mechanical laws could not explain a miracle: the fact that nature always produces the same crystals and the same chemical compounds, with the same properties. Bohr called this miracle the

[1]Revolving electrons implied the electrodynamic instability of Rutherford's model; something of which Rutherford was fully aware of. See section I.1.4 (point v).

> "stability of matter." It meant that an iron atom, after undergoing several modifications under external influences, always remains an iron atom with the same properties. It also meant that the light produced in a fluorescent vessel, filled with a particular gas, always presents the same spectral lines. Bohr noticed that this extraordinary stability could not be understood in the light of Newton's causal determinism, i.e., by the fact that the state of a system at a particular time should always be determined by the state of that system at an earlier time, and by nothing else ... To explain the existence of stable atoms, Bohr postulated the existence of "stationary atomic states."

In Bohr's paper of July 1913, the atom's stability was ensured by the presence of a stationary state of lowest energy, the "ground state". We will dwell in later sections on the reactions to Bohr's ideas. Let us first see how this young Danish physicist became interested in the Rutherford atomic model.

I.4.2 Bohr's first months in Manchester

Here are Bohr's reminiscences of his first meeting with Rutherford ([Bohr (1961)], p. 31):

> The first time I had the great experience of seeing and listening to Rutherford was in the autumn of 1911 when, after my university studies in Copenhagen, I was working with J. J. Thomson, and Rutherford came down from Manchester to speak at the annual Cavendish Dinner ... A few weeks after the Cavendish Dinner I went up to Manchester to visit one of my recently deceased father's colleagues who was also a close friend of Rutherford. There I again had the opportunity to see Rutherford, who in the meantime had attended the inaugural meeting of the Solvay Council in Brussels, where he had met Planck and Einstein for the first time. During the conversation, in which Rutherford spoke with characteristic enthusiasm about the many new prospects in physical science, he kindly assented to my wish to join the group working in his laboratory, when, in the early spring of 1912, I should have finished my studies in Cambridge, where I had been deeply interested in J. J. Thomson's original ideas on the constitution of atoms.

J. J. Thomson (1856–1940)
Courtesy of the Solvay Institutes

We do not know if in this conversation, Rutherford mentioned the Solvay discussions on the Haas model, and the possibility of linking atomic dimensions to Planck's quantum of action. As pointed out by Heilbron and Kuhn ([Heilbron and Kuhn (1969)], p. 238, see footnote 40), Bohr had apparently no knowledge of the work of Haas when he joined the Rutherford laboratory. During his first months in Manchester, "he [Bohr] displayed no interest whatsoever in radioactivity and in atomic models. His main concern was the electron theory of metals, subject of his doctoral thesis." However, a few months later "Bohr's main interests had shifted from the electron theory to matters of direct concern to the Rutherford group".

Bohr's transition to the problem of the atom took place in June 1912, as witnessed by this message to his brother Harald[2]:

> A couple of days ago, I had a little idea for understanding the absorption of α particles ... and I have worked out a little theory about it, which even if it is not much in itself, can perhaps shed a little light on some things concerning the structure of atoms. I am thinking of publishing a little paper about it very soon... In recent years (prof. Rutherford) has been working out a theory of atomic structure, which seems to be quite a bit more solidly based than anything else we've had before. And not that my work is of the same importance of kind, yet my result does not so badly agree with his.

Rutherford reacted favourably to Bohr's new interests, as shown by his message of June 12, 1912 to Martin Knudsen, the Secretary of ISIP's Scientific Committee[3]: "I have one of your countrymen, Dr. Bohr, working with me for a time. He seems a very able young fellow."

A week later, Bohr had more news for his brother ([Heilbron and Kuhn (1969)], p. 308):

> It could be that I've perhaps found out a little bit about the structure of atoms. You must not tell anyone anything about it; otherwise I certainly could not write you this soon. If I'm right, it would not be the indication of a possibility, but perhaps a little piece of reality. It has all grown out of a little piece of information I obtained from the absorption of α-particles (the little theory I wrote you about in my last). You understand that I still could be wrong, for it's not yet completely worked out (I believe it's not, however); nor do I believe that Rutherford thinks it's completely mad; but he is the right kind of man and would never say that he was convinced of something that was not entirely worked out.

As noted by Heilbron ([Heilbron (1985)], p. 306), the "little piece of information" to which Bohr refers may well have been his discovery that Rutherford's nuclear atom was also mechanically unstable[4]:

---

[2]Letter from Niels to Harald of June 12, 1912. See [Heilbron and Kuhn (1969)], p. 237.

[3]Archives de l'ESPCI, Fonds Langevin, L10/58.

[4]Besides the electromagnetic instability due to the radiation emitted by the accelerated electrons revolving around the nucleus, every planetary model also suffers from

> J. J. Thomson and the Cambridge School had rejected the nuclear model on account of its mechanical instability. Bohr welcomed it precisely because it needed a non-mechanical force to exist. Already in his Copenhagen dissertation he had pointed to certain phenomena – heat radiation and paramagnetism in particular – that eluded the electron theory and appeared to require the ascription of a non-mechanical rigidity to the paths of atomic electrons. He was drawn to the nuclear model as a possible representation or rectification of the sorts of difficulties he had encountered in his earlier studies. He saw in the quantum of action the "extra-mechanical condition" that ensured stability of the electronic orbits in a nuclear atom.

Bohr's confidence in Planck's quantum as a "stabilising factor" may have been strengthened by the available literature, notably the Solvay reports[5] [Solvay I (1912)] and Poincaré's papers [Poincaré (1911, 1912)], two significant sources published in the aftermath of the first Physics Council. The impact of Solvay I has been described by Heilbron and Kuhn as follows ([Heilbron and Kuhn (1969)], note 85 on p. 244):

> Before the first Solvay Congress very few people shared the conviction that Bohr had expressed in his thesis about the inevitability of a quantum-like break with classical mechanics. The Congress and Poincaré's response to it, did much to spread the view that Bohr had arrived at for himself. At the very least Bohr must have been reassured. Perhaps more important, Rutherford had been at the Congress and discussed it with Bohr...

Another factor may have contributed to Bohr's interest in the reports of Solvay I: the work of John W. Nicholson. As pointed out by Heilbron, J. ([Heilbron (1966)], note 18 on p. 340), it was under influence of the Council's discussions that Nicholson computed the angular momentum of

---

mechanical instability, which is due to the gravitational interaction among the constituents of the atom.

[5]As already mentioned, in the first part of his famous trilogy *On the Constitution of Atoms and Molecules*, which appeared in the *Philosophical Magazine* in July 1913, Bohr explicitly referred to the proceedings of Solvay I, published by Gauthier-Villars in January 1912.

his electrons, and found them to be integral multiples of Planck's constant. As noticed by Russell McCormmach ([McCormmach (1966)], note 54 on p. 170), in his papers, Nicholson also refers to Poincaré's communication [Poincaré (1912)], and to Sommerfeld's suggestion that the quantum of action should have an atomic significance.

I.4.3 Genesis of the Bohr atom

In his report to the twelfth Physics Council,[6] Bohr summarised the steps that led to his spectacular results for the spectrum of hydrogen-like atoms and the Rydberg constant:

> A starting point was offered by the empirical regularities exhibited by the optical spectra of the elements, which, as first recognised by Rydberg, could be expressed by the combination principle, according to which the frequency of any spectral line was represented with extreme accuracy as the difference between two members of a set of terms characteristic for the element. Leaning directly on Einstein's treatment of the photo-effect, it was in fact possible to interpret the combination law as evidence of elementary processes in which the atom under emission or absorption of monochromatic radiation was transferred from one to another of the so-called stationary states of the atom. This view, which permitted the product of Planck's constant and any of the spectral terms to be identified with the binding energy of the electron in the corresponding stationary state, also offered a simple explanation of the apparently capricious relationship between emission and absorption lines in series spectra, since in the former we are confronted with transitions from an excited state of the atom to some state of lower energy, while in the latter we generally have to do with a transition process from the ground state with the lowest energy to one of the excited states. Provisionally picturing such states of the electron system as planetary motions obeying Keplerian laws, it was found possible to deduce the Rydberg constant by suitable comparison with Planck's original expression for the energy states for a harmonic oscillator.

[6] See [Solvay XII (1962)], pp. 18–19.

> The intimate relation with Rutherford's atomic model appeared not least in the simple relationship between the spectrum of the hydrogen atom and that of the helium atom, in which one has to do with systems consisting of an electron bound to a nucleus of minute extension and carrying one and two elementary electric charges, respectively.

Despite the above resounding successes, the publication in July 1913 of Bohr's first paper on the constitution of atoms aroused relatively little interest. However, a striking reaction was that of Ehrenfest, Lorentz's successor at the University of Leiden. In a letter to Lorentz of August 25, 1913, he made this devastating comment[7]: "Bohr's work on the quantum theory of the Balmer formula (in the *Philosophical Magazine* journal) has driven me to despair. If this path leads to the goal, I must give up physics."

Henry G.J. Moseley (1887–1915)
Courtesy of Smithsonian Institution Archives,
Acc. 90-105, Image no. SIA2009-1116

[7]See [Kragh (2010)], p. 9 and [Kox (2018)], p. 368.

Moseley, on the contrary, was quite positive about Bohr's work, even if remaining cautious. According to Kragh ([Kragh (2010)], p. 18):

> Together with Rutherford, Moseley was the most important of the early supporters of Bohr's atomic theory. His experimental results became the best advertisement for the Bohr-Rutherford nuclear atom. Moseley reported his first series of experiments on X-ray spectroscopy in a letter to Bohr of November 16, 1913, saying "the results lend great weight to the general principles which you use." On the other hand, Moseley was careful not to link his work too closely to the Bohr atom. In his famous paper of 1914, including the first version of the Moseley diagram, he did not refer to Bohr's theory, but only concluded from the evidence of the X-ray spectra alone, without using any theory of atomic structure, that the chemical elements must be characterised by an integer, that is the atomic number.

In his report to the twelfth Physics Council, Bohr recognised the decisive support to his atomic theory that had come from the studies of X-rays, carried out by Moseley at the time of Solvay II, and extended in 1914. As we shall see in section I.6.4, Moseley was one of the beneficiaries of Solvay subsidies. Here is the tribute that Bohr paid in his Rutherford Memorial Lecture to this brilliant British physicist, whose career had been shattered by the war ([Bohr (1961)], p. 44):

> In modern history of physics and chemistry, few events have from the outset attracted such general interest as Moseley's discovery of the simple laws allowing the unambiguous assignment of the atomic number to any element from its high-frequency spectrum. Not only was the decisive support of the Rutherford atomic model immediately recognised, but also the intuition which had led Dmitri Mendeleev at certain place in his table to depart from the sequence of increasing atomic weights was strikingly brought out. In particular, it was evident that Moseley's laws offered an unerring guide in the search for as yet undiscovered elements fitting into vacant places in the series of atomic numbers.

Bohr never failed to emphasise the change that Moseley's work of 1913–1914 had brought about in the perception of the Rutherford–Bohr atomic model ([Heilbron (1977)], p. 74): "The Rutherford work was (at first) not taken seriously ... There was no mention of it in any place. The great change came from Moseley."

Henry G.J. Moseley (1887–1915)
Courtesy of Smithsonian Institution Archives, Acc. 90-105, Image no. SIA2009-1115

Chapter I.5

Solvay II (1913): The structure of matter

Solvay II (official picture) – October 27–31, 1913
Courtesy of the Solvay Institutes

Details on reports and discussions of the second Physics Council can be found in the volume published by Gauthier-Villars [Solvay II (1921)]. Again we limit ourselves to reporting some facts that shed light on the purpose of our book. We rely on *first hand* sources, such as the *Notes* written by Rutherford and collected by Maurice de Broglie during the Solvay meeting [M. de Broglie (1911)], and the report presented by Bohr at the twelfth Physics Council [Solvay XII (1962)]. We keep making use of more recent analyses, made by leading historians of science.

Another picture taken during Solvay II The Council was presumably discussing Laue's discovery: on the blackboard, on the right, Laue's spots are represented as in ([Solvay II (1921)], p. 80).
Courtesy of the Solvay Institutes

I.5.1 Bohr about Solvay II

Bohr, as we know, did not take part in the second Physics Council. However, at the time of the conference his first paper on the constitution of atoms had been in print for three months. Here are the lines that Bohr devoted to Solvay II ([Solvay XII (1962)], p. 17):

> By the time of the next Solvay conference in 1913, the subject of which was *the Structure of Matter*, most important new information had been obtained by Laue's discovery in 1912 of the diffraction of Röntgen rays in crystals. The discovery removed indeed all doubts about the necessity of ascribing wave-properties to this penetrating radiation, the corpuscular features of which in its interaction with matter, as especially stressed by William Bragg, had been so strikingly illustrated by Wilson's cloud chamber pictures showing the tracks of high speed electrons liberated by the absorption of the radiation in gases.
>
> As is well known, Laue's discovery was the direct incentive to the brilliant explorations of crystalline structure by William Lawrence Bragg, who by analysing the reflection of monochromatic radiation from the various sequences of parallel plane configurations of atoms in crystal lattices were able both to determine the wavelength of the radiation and deduce the type of symmetry of the lattice.
>
> The discussion of these developments, which formed the main topic of the conference, was preceded by a report by J. J. Thomson about the ingenious conceptions regarding the electronic constitution of atoms, by which, without departing from classical physical principles, he had been able, at least in a qualitative way, to explore many general properties of matter. It is illuminating for the understanding of the general attitude of physicists at that time that the uniqueness of the fundament for such exploration given by Rutherford's discovery of the atomic nucleus was not yet generally appreciated.
>
> The only reference to this discovery was made by Rutherford himself, who in the discussion following Thomson's report insisted on the abundance and accuracy of the experimental evidence underlying the nuclear model of the atom. Actually, a few months before the conference my first paper on the quantum theory of atomic constitution had been published, in which initial steps had been taken to use the Rutherford atomic model for the explanation

> of specific properties of the elements, depending on the binding of the electrons surrounding the nucleus. As already indicated, this question presented unsurmountable difficulties when treated on ordinary ideas of mechanics and electrodynamics, according to which no system of point charges admits of stable equilibrium, and any motion of the electrons around the nucleus would give rise to a dissipation off energy through electromagnetic radiation, accompanied by a rapid contraction of the electron orbits into a neutral system far smaller than the size of atoms derived from general physical and chemical experience.
>
> This situation therefore suggested that the treatment of the stability problems be based directly on the individual character of the atomic processes by the discovery of the quantum of action.

We notice that in this part of his report, Bohr devotes only two lines to the historic debate that followed the presentation of Thomson's report, stating that Rutherford "insisted on the abundance and accuracy of the experimental evidence underlying the nuclear model of the atom". In fact, we know from the Council's Minutes that Thomson referred in his report to the "great and abrupt deviations" observed in the Manchester α scattering experiments, but with the claim that they did not constitute evidence in support of Rutherford's hypothesis. Thomson's main argument was that an atomic nucleus, of the kind proposed by Rutherford, should produce deviations in β-rays similar to the ones observed in α-rays, and that no such effect had ever been recorded in a Wilson cloud chamber. We also know, thanks to the Notes collected by de Broglie, that Rutherford challenged Thomson's argument during the discussion of his report, delivering a first documented plea in favour of an atomic nucleus before an international audience.[1] These were his words:

> Sir J. J. Thomson has drawn our attention to the salient fact that the trails of the β particles recorded photographically by C. T. R. Wilson do not show the number of large deflexions to be expected from the theory of the atomic nucleus which I put

[1]Rutherford had brought the idea up at the British Association for the Advancement of Science (BAAS) meeting of September 1913, but without discussing the experimental data on which it was based (a brief report of the meeting has been published in *Nature* [BAAS (1913)]).

forward some time ago. I would like, first of all, to emphasise the definiteness of the experimental evidence on which the theory is based. It is well known that α particles are occasionally deflected through a large angle and the evidence indicates that this is the result not of a multitude of atomic encounters but of a close encounter with a single atom. When we consider the great velocity and energy of the α particles, we are faced to the conclusion that in order to suffer such a large deviation, they must pass through an intense electric field within the atom. In order to account for these results it was suggested that the atom consists of a central nucleus of exceedingly small dimensions, carrying a positive charge and containing most of the mass of the atom, surrounded by a distribution of electrons to render the atom electrically neutral. The large deflexions were assumed to be due to the passage of α particles through the intense field close to the central nucleus of an atom. A careful comparison of the theory with experiment was made by Geiger and Marsden, and the deductions drawn from the theory were shown to be in excellent accord with the experimental results. It was deduced that the charge on the nucleus, and consequently the number of electrons in the atom, was about equal to half the atomic weight for all the elements examined. The force between the α particle and the nucleus was assumed to follow the ordinary inverse square law, and it was found that any other law was inconsistent with the experimental facts. It could naturally be anticipated that high velocity electrons of energy about the same as the α particles should also suffer large single deflexions in passing close to the nucleus. It must be remembered, however, that while the energy of the β particle considered by Professor Thomson is much smaller than that of the α particles, it is supposed to pass within the strong field due to the central nucleus. I think it is doubtful whether the slow velocity β particle penetrates at all the region considered where the α particle suffers large deflexions, but is more likely to be confined to the region of the atom where the field due to the nucleus is much weakened, and where it suffers small deflexions due to encounters with the electrons associated with the atom. The question whether a high velocity β particle, of energy comparable to that of the α particle, suffers large simple

> deflexions is of great theoretical importance, and is now being examined experimentally. Unless one supposes that the atoms have a charged nucleus of small dimension, it is impossible to explain the experimental facts of scattering of α particles without assuming that new and unsuspected forces of attraction or repulsion of an intense kind exist between atoms when close together. Considering that the nucleus theory gives the same number of electrons in an atom as the scattering of X-rays, it seems to me simpler to suppose that such a charged nucleus has an actual existence and is a fundamental constituent of all atoms.

Rutherford's point of view was immediately shared by Langevin, who made the following comment [Solvay II (1921)]

> Measurements of deviations of α-particles led Mr. Rutherford to suppose that the central positive nucleus of the atom is surrounded by a number of *peripheral electrons*[2] approximately equal to half the atomic weight, that is to say approximately equal to the total number of electrons present in the atom, according to the result of measurements of dispersion of Röntgen rays. Can this be reconciled with the idea that radioactive phenomena have their seat in this central core? The nucleus should contain the electrons emitted in the form of β-particles during *radioactive transformations*.[3] It is difficult not to admit for these β-rays an origin deep inside the atom. [emphasis added]

Marie Curie agreed, but insisted that it was only to peripheral electrons that one should apply the result according to which the number of electrons per atom is proportional to the atomic weight, and approximately equal to half this number. As for the assumed "essential electrons", deep inside the atom, she prudently declared: "We still don't know anything about their number."

---

[2]The notion of peripheral electrons, to be distinguished from those believed to be responsible for radioactive phenomena, had been put forward by Marie Curie at Solvay I, and was reaffirmed by her at Solvay II during the discussion of Thomson's report.

[3]At the time, Marie Curie, Langevin and many others, regarded the emission of β-particles with extremely high velocities, as an indication that the atomic nucleus should contain essential electrons (according to a terminology introduced by Marie Curie at Solvay II). This belief would only be abandoned after the discovery of the neutron (1932), following Heisenberg's report at the sixth Physics Council in 1933. See reference [Solvay VI (1928)].

Maurice de Broglie, for his part, did not limit himself to collecting remarks made in German or English, as he had done at Solvay I. He took an active part in the discussion of several reports, starting with that of Thomson. Thus, following Marie Curie's remark about the distinction that should be made between peripheral atomic electrons and those that are emitted as β-particles, he declared [Solvay II (1921)]: "Even in the case of external electrons, we must renounce ordinary mechanics if we admit quanta in photoelectric phenomena." De Broglie's statement leaves little doubt about his belief that quanta should play a major role in the mechanism at work in photoelectric processes.

I.5.2 Rutherford's silence on Bohr's atomic theory

We saw that, in his report, Bohr mentioned the publication in July 1913 of the first part of his famous "trilogy" [Bohr (1913b, c, d)]. However, he did not say a word about the fact that neither Thomson, nor Rutherford, deemed it necessary to refer to his theory. In fact, Thomson quoted Bohr's name, but for an earlier work.[4] As for Rutherford, he named Bohr in a statement regarding the number of atomic electrons:

> It is difficult to fix with certainty, by the present methods, the actual number of electrons in a heavy atom. It may well be, as Bohr has suggested, that this number is equal to the number of the atom when the atoms are arranged on a series in order of their atomic weights.

Rutherford's silence on the first part of Bohr's trilogy (a paper sent by Rutherford on July 1913 for publication to the *Philosophical Magazine*), is all the more surprising since its content had been favourably commented by Jeans at the Birmingham meeting of BAAS held in September 1913, i.e. before Solvay II ([Kragh (2010)], p. 12).[5] In the first part of his trilogy, Bohr did more than provide proof that the Rutherford atom could account for the specific properties of the chemical elements; he showed that it explained specific facts discussed at Solvay II, notably Einstein's photoelectric law and Wood's experimental results on the absorption of light by sodium vapour ([Kragh (2010)], p. 7).

---

[4]Thomson, mentioned Bohr's analysis of the collisions between charged particles and atomic electrons in circular orbital motion [Bohr (1913a)].

[5]This meeting had been attended by Bohr and by six members of Solvay II: Lorentz, Rutherford, Marie Curie, Thomson, W. H. Bragg, and Wood. See p. 379 of the Report of the British Association for the Advancement of Science.

Two elements can be given to explain Rutherford's attitude. One was the fact he did not need Bohr's revolutionary ideas to refute J. J. Thomson's ill-founded claim. Another was Rutherford's "uneasiness" about Bohr's theoretical views. In a letter written after reading the manuscript of the paper of July 1913, and dated March 20, 1913, Rutherford told Bohr[6]: "It is very difficult to form a physical idea of the basis of your theory."

Rutherford's ambiguous position with respect to Bohr's theory was commented on by Helge Kragh ([Kragh (2010)], pp. 17–21):

> Rutherford, who had closely followed the development of Bohr's ideas of atomic structure, was of course much in favour of the new atomic theory which complemented and justified his own earlier theory of the nuclear atom. But Rutherford's research interest was radioactivity and the atomic nucleus, not the electron system with which Bohr's theory was primarily about ... Although advocating Bohr's theory, Rutherford was somewhat cautious and reluctant to comment on the central parts of the theory, dealing with spectroscopy and the quantum postulates.

Whatever Rutherford's main motives for remaining silent about Bohr's atomic theory, we have reasons to believe that Bohr's work was discussed in private during the meeting. We keep quoting Kragh:

> Arnold Sommerfeld was among the physicists to whom Bohr sent preprints of his paper in the July issue of the *Philosophical Magazine*. In a postcard of September 4, 1913, he told Bohr: "I thank you very much for sending me your highly interesting work ... Though for the present I am still rather sceptical about atomic models in general, calculating the (Rydberg) constant is undoubtedly a great feat ... From Mr. Rutherford, whom I hope to see in October (at the Solvay meeting), I may perhaps learn more about your plans."

We do not know whether Rutherford spoke with Sommerfeld about Bohr's theory, nor what he said, but it is likely that once again Rutherford remained extremely cautious.[7] His uneasiness with theoretical speculation was intimately linked with his permanent wish to focus on experimental

[6]Rutherford's letter to Bohr was dated March 20, 1913.
[7]See section I.1.4.5.

evidence. This fact will be confirmed in section I.6.4 by the importance given to Moseley's work.

I.5.3 Marie Curie on intra-atomic disorder

During the discussion of Thomson's report, Marie Curie presented a Note "On the fundamental law of radioactive transformations." The Note contained ideas of her collaborator André Debierne[8] (the discoverer of actinium) about the information that the law of radioactive transformations might provide on the constitution of the inner part of atoms.

Marie Curie pointed out that an interpretation of this law,[9] which applies to all radioactive substances, arises immediately if one admits that the transformation is determined by a cause external to the atom. Since experience showed that temperature had no influence on radioactive transformations, it seemed natural to look for an *element of disorder*, inside the atom itself, that might explain the appearance of a law of chance. Expanding on Debierne's ideas, Marie Curie suggested that each atom passes, during a small interval of time, through a large number of different states, the average state remaining constant and independent of external factors. To achieve this succession of variable states, with a constant probability of destruction, Debierne had been led to admit the presence inside the atom of a large number of elements endowed with disordered movements. Their agitation inside the atomic core could produce, from time to time, an *unstable* state that would lead to the transformation of the atom. Marie Curie remarked, for her part, that this intra-atomic disorder could arise in the case of a fairly simple structure, the large number of atoms being responsible for the appearance of the law of chance. This consideration was supported by Langevin. He referred to the complexity of the trajectories obtained by Poincaré in a case as simple as that of the three-body problem, and concluded:

> We would not need to greatly increase the number of degrees of freedom of the system in order to make sure that each atom

---

[8]Debierne cherished the hope of being invited to present his Note. Unable to respond to this request, Lorentz suggested that the Note, which he considered worthy of interest, be presented by Marie Curie (see *Procès-verbaux des réunions du Comité scientifique Solvay*, Archives de l'ESPCI, Fonds Langevin).

[9]The law of radioactive transformations is the experimental observation that the probability of the transformation of an atom, at a given moment, is the same for all atoms.

> passes, over time, from an initial configuration as closely as needed to a given configuration, for example one that allows spontaneous destruction.

Debierne's suggestion, gave rise to various reactions. Rutherford defended the view that new and important knowledge on the constitution of atoms should come from the study of radioactive transformations. He agreed, in particular, with Debierne's claim that the universal law of radioactive transformation should be seen as a consequence of fortuitous disturbances occurring in the atomic nucleus. However, remaining cautious, he declared: "In the current state of our knowledge, it does not seem possible to get a clear idea as to the constitution of the atomic nucleus itself, nor of the causes that lead to its disintegration".

In a communication to *Nature*, published after Solvay II, on December 6, 1913, Rutherford made the following comments on a paper by Fredrick Soddy on the bearing of his theory on radioactive phenomena ([Rutherford (1913b)], p. 423):

> Mr. Soddy ... seems to be under the impression that I hold the view that the nucleus must consist entirely of positive electricity. As a matter of fact, I have not discussed in any detail the question of the constitution of the nucleus beyond the statement that it must have a resultant positive charge.
>
> There appears to me no doubt that the α particle does arise from the nucleus, and I have thought for some time that the evidence points to the conclusion that the β particle has a similar origin.
>
> This point has been discussed ... by Bohr ... The strongest evidence in support of this view is, to my mind, (1) that the β ray – like α ray – transformations are independent of physical and chemical conditions, and (2) that the energy emitted in the form of β and γ rays by the transformation of an atom of radium C is much greater than could be expected to be stored up in the external electronic system ...

Rutherford's statements confirm his caution, and his constant concern to favour experimental observation rather than theoretical conjecture.

I.5.4 Einstein on "zero-point energy"

The following two points concern Einstein's reaction to Bragg's report, and his admission of being wrong about a long debated question at Solvay I. Here is an excerpt of Einstein's letter to Ludwig Hopf of November 2, 1913 ([Klein *et al.* (1994b)], doc. 480):

> In Brussels the lecture of Bragg Sr. was extremely interesting. It is unbelievable how much this man has already found out about the lattice structure of crystals and about Röntgen rays. Now, all at once, it is possible to carry out exact determinations of the wavelengths of Röntgen rays. Some metals (e.g. radium) display very narrow emission ranges, so that it is really possible to produce monochromatic Röntgen rays. One can safely say that Laue's figures are totally explained.

On the other hand, Einstein intervened twice during the discussion of Laue's report about the question of *zero-point energy*, a prediction of Planck's emission theory. After pointing at the fact that a persistence at absolute zero of elastic atomic vibrations (i.e. thermal energy) was in contradiction with the phenomenon of *superconductivity*, observed by Kamerlingh Onnes, he declared "invalid" the arguments he and Otto Stern [Einstein and Stern (1912)] had put forward in a recent paper on zero-point energy.

I.5.5 Sommerfeld's clarification of X-ray diffraction

Among the significant contributions to Solvay II, there is one which deserves special attention: the note entitled[10] *On the Quaternary and Ternary Photograms of the Blende, and the Spectrum of Röntgen Radiation*, presented by Sommerfeld during the discussion of W. H. Bragg's report.

As recognised by Sommerfeld in a letter to his wife,[11] his unscheduled contribution was meant to provide a complement to Laue's report:

> The day's work is now over: sessions from 9 a.m. till 1 p.m., lunch until 4 p.m., followed by a road trip to Laeken, to

[10]The German title reads: *Über die vierzähligen und dreizahligen Photogramme der Zinkblende und das Spektrum der Röntgen–Strahlung.*

[11]One of the authors (F.L.) wishes to thank Mrs. Monika Baier, a granddaughter of Sommerfeld, for providing him with a copy of this letter.

> visit a wireless telegraphy station. This morning I brought my contribution to the Laue phenomenon; it was good overall, but poorly presented (even if Einstein does not think so), because I did not sleep well. From now on, I will be a listener...

The truth is that by focusing on the special case of a Zinc Sulphide (ZnS) crystal, and by comparing Laue's interpretation of X-ray diffraction to the one provided by W. L. Bragg, Sommerfeld lifted the remaining uncertainties about the link between these two interpretations. We quote Eckert ([Eckert (2015)], p. 2072):

> At the time of this Solvay Congress it was not yet quite clear how Laue's theory of three-dimensional diffraction and the concept of Bragg reflection from crystal planes were related to each other. Sommerfeld's presentation was dedicated to the proof that both yield the same results. He chose a special case (ZnS) in order to demonstrate how the crystal selects from the continuous spectrum of the primary X-ray Bremsstrahlung the monochromatic wavelengths that are observed in the interference pattern. Looking back, the main result of this work was the full confirmation of W. L. Bragg's structure determination and of his physical picture of the pass filter action of the crystal.

The importance of Sommerfeld's contribution was recognised by Paul Ewald ([Ewald (1962)], p. 78).[12] Today, we can say that his analysis shaped the modern interpretation of X-ray diffraction. At Solvay II, Lorentz was struck by the importance of Sommerfeld's unscheduled "note", to the point of presenting it in the Council's Minutes as a genuine Solvay report. He did the same with Marie Curie's Note entitled *On the fundamental law of radioactive transformations.*

I.5.6 Impact of Wood's report

In his report on resonance radiation, Wood presented resonance spectra, the complexity of which seemed to defy theoretical interpretation. As pointed out by M. de Broglie ([M. de Broglie (1951)], p. 39), these spectra would later conform to the patterns drawn from Bohr's atomic theory and its improvements.

---

[12] Sommerfeld acknowledged in Brussels Ewald's collaboration on this question.

During his presentation, Wood announced that he and Louis Dunoyer, a French expert in optical instruments, intended to carry out experiments on "selective reflection, scattering and absorption of radiation by resonating gas molecules". He mentioned a paper on the subject that had recently been published by Dunoyer [Dunoyer (1913)]. As we already noted, Wood's report had a direct effect on the decisions taken by the ISC with regard to the granting of subsidies. We will see that the Committee ranked Dunoyer in fifth position (i.e. well above Moseley) on the list of subsidies for the budget year 1913–1914, and that it decided to grant him of 2,700 francs for a "Study of optical resonance".

Chapter I.6

Solvay grants: November 1912–July 1914

Sir Ernest Rutherford
Courtesy of the Solvay Institutes

Marie Skłodowska-Curie
Courtesy of the Solvay Institutes

Two notable members of the International Scientific Committee (ISC). They spared no effort in examining the research projects submitted for funding to ISIP, and took the trouble to write reports.

I.6.1 Decisions taken on the sidelines of Solvay II

Four ISC meetings took place from October 27 to October 31, 1913. The first point on the agenda was the publication of the reports presented during the last Council. Some Committee members defended the idea of a volume similar to the one of 1912 (which had been published in French out of courtesy to Mr. Solvay). Others argued that all contributions should be published in their original form. The Committee finally decided in favour of this last option, and agreed on preparing an account of the discussions before November 15. Goldschmidt was asked to take care of the printing of reports and discussions.[1]

A multilingual volume with the reports and discussions of Solvay II, was edited under Lorentz's supervision, and printed in 1916 at the cost of great effort.[2] Unfortunately, its publication was suspended, due to the country's occupation. Almost all existing copies were destroyed after the war by decision of ISIP's Administrative Commission.[3] After the Great War, the option of a multilingual report was abandoned. An official volume, with the reports and discussions of Solvay II translated in French, was published by Gauthier-Villars in 1921.

Regarding the granting of subsidies, a first round of discussions took place in the evening of October 30. Impressed by the advances realised in the domain of X-ray diffraction, Lorentz proposed to provide additional support to Laue, and to grant a subsidy to Bragg. Rutherford told his colleagues that Bragg senior had no intention of asking for help. Hence, Lorentz suggested to find out if Laue needed additional means to continue his research in Zurich.

The Committee decided to examine the applications that had been submitted before September 15, 1913. A first evaluation of the projects had been carried out before the meeting. Each project had been examined by a member of the Committee, and had been the subject of a report. The reports had been sent to the other members, who had been asked to make comments and establish a list of priorities. A final priority list, based on these data, had been drawn up by Lorentz. Many requests were rejected on the basis of poor quality or inappropriate content. The successful ones

[1]The job was entrusted to the printing house Hayez, the official publisher of Belgium's Royal Academy.

[2]See [Lambert and Berends (2021)], section 8.4.

[3]Two copies of the Hayez volume have been preserved. They are kept at the *Service des Bibliothèques et Archives de l'Université Libre de Bruxelles* (S.a.b.ULB).

were divided into two groups. A first one, for the budget year 1912–1913, comprised 15 projects submitted before February 15, 1913. A second list comprised 13 projects submitted between February 15 and September 15, 1913. The latter would be financed with the budgetary means for the year 1913–1914.

I.6.2 Noteworthy beneficiaries of Solvay grants

From the full list of beneficiaries,[4] we selected eleven *noteworthy cases.* They involve applications submitted by physicists whose names are cited by Bohr in his Rutherford Memorial Lecture [Bohr (1961)].

Other less-known names of successful applicants could also be singled out for having been cited for their work in Solvay reports. As an example, we briefly consider the case of Harry Dember, a German researcher whose work on X-rays was cited by Sommerfeld at Solvay I. Dember was granted a significant subsidy for the following reason. In an attempt to determine the lower limit of the potential differences under which X-rays could be produced, he noticed that "very soft rays" were observed by the stopping of cathode rays emitted under 250 Volts. Ranked in eighth position on the priority list for the year 1912–1913, Dember received 1,500 francs for *Research on canal rays and on Röntgen rays produced in the photoelectric effect.* We now examine the projects submitted by successful applicants, cited for their work by Bohr.

I.6.3 Subsidies for the year 1912–1913

Among the fifteen subsidies granted for projects submitted before February 15, 1913 ([Lambert and Berends (2021)], p. 168), we chose three cases. They are indicated in order of priority, with the name of the applicant, the amount granted, the name of the referee, and the amount requested. The order was established by Lorentz "on the basis of individual evaluation supplied to him by the ISC members" ([Lambert and Berends (2021)], p. 168):

- In first position: **Sommerfeld**, 4,000 francs for *Continuation of Mr. Friedrich's research on Röntgen rays* (report by Rutherford; sum requested: 5,000 francs).

[4] A complete list of beneficiaries of Solvay grants, before the outbreak of WWI, can be found in [Lambert and Berends (2021)], pp. 172–173.

- In seventh position: Edgar **Meyer**, 3,000 francs for *Research on the nature of γ rays* (report by Marie Curie; sum requested: 3,435 francs).
- In tenth position: James **Franck and** G. **Hertz**, 2,000 francs for *Research on the ionisation work and free path of electrons in gases* (report by Knudsen; sum requested: 2,500 francs).

Before commenting on these cases we wish to underline Marie Curie's and Rutherford's acceptance to examine the many projects submitted to ISIP during the years 1913–1914. Rutherford's willingness to report on some of them should be specially praised, since he was known for not liking to write reports ([Badash (1979)], p. 34).

Sommerfeld

Two factors may have played a role in the high ranking of Sommerfeld's request. One was his brilliant presentation on October 29, the day before the ISC meeting. The other was the Committee's wish to provide the Munich laboratory with the means that would allow Friedrich to continue his experiments after Laue's departure. This is the message that Sommerfeld sent to Lorentz[5] on January 14, 1913:

> I am intentionally submitting this application only now, after having been informed that the corresponding application by colleague Laue has been approved, so not compete with him. Mr. Laue, had to reorganise himself in Zurich, and was therefore in greater need, while I am able to finance the continuation of the experiments we have agreed on, partly from the funds of my institute. However, the ideal X-ray apparatus which I have been granted for the time being, already exceeds these means by far. I hope to be able to ensure that the requested sum will bear good fruit in my institute with the enclosed application.

Meyer

Edgar Meyer, a former "Privatdozent" in Aachen, was an expert in the study of γ-rays. At Solvay I, Sommerfeld and Einstein pointed at the importance of his experiments on the fluctuations in the ionisation current produced by γ-rays. According to Wheaton ([Wheaton (1983)], p. 150), it was under the influence of Meyer's results on these fluctuations that

[5]Haarlem, Noord-Hollands Archief, Archief Lorentz, inv. num. 74. We thank Michael Eckert for providing us with a copy of this letter.

Sommerfeld "pushed his pulse theory of γ-rays to the extremes, to describe the radiation produced by β-electrons accelerated out of an atom at a velocity close to that of light."[6] There are also indications that Meyer's recourse to a statistical theory of radioactive decay played a role in Planck's introduction of statistical elements in his second quantum theory ([Wheaton (1983)], p. 179). We therefore assume that Marie Curie, who reported on Meyer's request, had no difficulty in convincing her colleagues of the need to support his project.

Franck & Hertz

James Franck and Gustav Hertz are famous for their experiment in which a beam of electrons was sent on mercury vapour. Their results, presented on April 24, 1914 to the German Physical Society, are generally cited as the earliest evidence in favour of Bohr's postulates. We quote Bohr ([Bohr (1961)], p. 48):

> As regards the problem of atomic constitution, a new impulse was given by the publication in 1914 of the famous experiments by Franck and Hertz on the excitation of atoms by electron impact.

However, when Franck and Hertz submitted their subsidy request, they had no idea of Bohr's work. We quote Kragh ([Kragh (2010)], p. 75):

> In an interview many years later, Franck recalled that he and Hertz were unaware of Bohr's theory when they made their key experiments in 1914: "We had neither read nor heard about it. We had not read it because we were negligent to read the literature well enough... On the other hand, one would think that other people would have told us about it. For instance, we had a colloquium at that time in Berlin at which all the important papers were discussed. Nobody discussed Bohr's theory." ... When the experiments nevertheless played a role in the reception of Bohr's theory, it was not due to the two German physicists, but to Bohr, who in 1915 pointed out how the data obtained in Berlin could be understood as a confirmation of his theory of stationary states.

Fortunately, Lorentz and his colleagues were wise enough to propose a grant of 2,000 francs for Franck and Hertz. It is likely that they were

[6] This theory was presented by at Solvay I.

encouraged in their decision by Warburg, the President of the Physikalisch-Technische Reichsanstalt, who during the discussion of J. J. Thomson's report, mentioned the ionisation results obtained by Franck and Hertz for oxygen and nitrogen ([Solvay II (1921)], p. 63).

There is an interesting point that should be added regarding the support given to Franck and Hertz. These physicists would have been deprived of a grant had there not been a most welcome and unexpected intervention on the part of Solvay. Indeed, the budgetary means for the year 1912-13 only allowed the support of 8 projects. Having learned that 7 other projects had been considered worthy of support, Solvay decided to intervene in a manner "that would allow the Institute to go a little further in granting subsidies than permitted by the available resources" [Lambert and Berends (2021)], chapter 7). Thanks to this new burst of generosity, the Institute found itself in a position to honour 15 requests, instead of 8.

Franck and Hertz received their grant at the end of 1913, and thanked the Institute in 1914 ([Frank and Hertz (1914)], p. 467). With hindsight, one can say that the poor ranking of their project (compared with that of Dember) is a striking illustration of the unpredictability of success in physics.

I.6.4 Subsidies for the year 1913–1914

The following subsidies were granted for projects submitted between February 15 and September 15, 1913:

- In fifth position: Louis **Dunoyer**, 2,700 francs for a *Study of optical resonance* (report by Brillouin; sum requested: 2,675 francs).
- In sixth position: Kasimir **Fajans**, 500 francs for *Research on the filiation of the elements* (report by Marie Curie; sum requested: 500 francs).
- In eleventh position: H. G. J. **Moseley**, 1,000 francs, for *Research on the reflection of X-rays* (report by Rutherford; sum requested: £50).

We examine again each case following the order of priorities fixed by the Committee.

Dunoyer

Dunoyer, the already cited expert in optics, is known today for having invented the *molecular beam method.* Developed by Otto Stern, this technique was used in the Stern-Gerlach experiment[7] of 1922, establishing the quantisation of the orientation of angular momentum. In October 1913, the merits of Dunoyer were highlighted, as we have seen, by Wood. It was apparently on the basis of his report that the ISC decided to support Dunoyer's research with a substantial grant.

Fajans

Fajans, a Polish chemist, had been working in 1910 at Rutherford's laboratory. As noted by Lawrence Badash ([Badash (1979)], p. 34):

> [Rutherford] recognised the need for chemical expertise, and was fortunate in his associations with Frederick Soddy, Bertram Boltwood and Otto Hahn at McGill, and Kasimir Fajans, [Alexander Stuart] A.S. Russel and George de Hevesy in Manchester.

In 1911, Fajans published a paper with Moseley on "Radioactive Products of Short Life".[8] In the beginning of 1913, when working at the Institute of Physical Chemistry of the Technische Hochschule in Karlsruhe, Fajans published a paper entitled "Radioactive Transformation and the Periodic System of the Elements", which appeared in the journal *Chemical News.* In fact, the law that governs the transformation of elements during radioactive decay is known today as the *Radioactive Displacement law of Fajans and Soddy.* We quote Bohr ([Bohr (1961)], p. 34):

> The radioactive displacement law in its complete form was enunciated in the beginning of 1913 by Soddy working in Glasgow, as well as by Fajans, working in Karlsruhe. However, these authors did not recognise its close relation to the fundamental features of Rutherford's atomic model, and Fajans even regarded the change in chemical properties, evidently connected with the electronic constitution of the atoms, as a strong argument against a model according to which the α- as well as the β-rays had their origin in the nucleus.

Fajans obtained 500 francs for his project, the minimum amount for a subsidy. Shortly afterwards, a prediction of Fajans and Soddy on the

[7]The experiment was set up by Otto Stern and Walther Gerlach.

[8]The paper appeared in number 48 of the *Berichte der Deutschen Gesellschaft.*

nature of radium B was corroborated by Rutherford. We quote Wheaton ([Wheaton (1983)], p. 225):

> At the end of 1913 Rutherford's data on the wavelengths of γ-ray lines produced unexpected corroboration of Kasimir Fajans and Soddy's prediction that radium B is an isotope of lead. This justification of the hypothesis of radioactive decay series encouraged Rutherford's belief that a smooth transition would be found between X-rays and γ-rays.

Moseley

When we think of the importance of Moseley's results on the "High Frequency Spectrum of the Elements", we can only be surprised at the low ranking of his project. The more so since it was Rutherford who wrote the report. As noted by Heilbron ([Heilbron (1966)], p. 339), Moseley had been an attentive witness of the steps that led to Rutherford's nuclear hypothesis:

> Within a few months after Moseley's arrival in Manchester, Rutherford had worked out his atomic model and scattering theory, which was published in the *Philosophical Magazine* for April 1911. Throughout the following two years the improvement of the theory and scattering data occupied several of the more advanced men at the laboratory, H. Geiger, C.G. Darwin, and N. Bohr. Moseley would not have missed the main point of this investigation, the effort to obtain precise and consistent agreement between the data and the theoretical quantities which chiefly determined the scattering: the number of electrons in the atom (N) and the charge on the nucleus (Ne).[9] Nor would other contemporary views of the atom have escaped his attention. For example, the deliberations of the first Solvay Congress, emphasising the importance of the quantum for the problem of atomic structure, might well have been described to him by Rutherford, who was one of the participants.

Many years later, Bohr recalled Moseley's presence at the Rutherford laboratory in June 1913, and the experiments he intended to carry out ([Bohr (1961)], p. 45):

[9] The symbol e stands for the electric charge of the *electron*.

> On my visit to Manchester in June, I had discussed with Darwin and Moseley the question of the proper sequence for the arrangement of the elements according to their atomic number, and learned then for the first time about Moseley's plans to settle this problem by systematic measurements of the high-frequency spectra of the elements by the Laue-Bragg method. With Moseley's extraordinary energy and gifts, of purposeful experimentation, his work developed astonishingly quickly, and already in November 1913, I received a most interesting letter from him with an account of his important results and with some questions regarding their interpretation on the lines which had proved applicable to the optical spectra.

On another occasion, Bohr insisted on the efforts made by Moseley at the time of Solvay II, and on the far-reaching consequences of his results[10]:

> It is of interest to recall that at the very time of the conference Moseley was studying the high frequency spectra of the elements by the Laue-Bragg method and had already discovered the remarkably simple laws which not only allowed the identification of the nuclear charge of any element, but even were to give the first direct indication of the shell-structure of the electronic configuration in the atom responsible for the peculiar periodicity exhibited in Mendeleev's famous table.

According to André Langevin ([Langevin (1966)], p. 7), his father – Paul Langevin – received a letter signed by Moseley on September 10, 1913, in which the British researcher asked for money to pursue his X-ray research on crystals. As far as we know, the letter got lost. André Langevin suggests that his father played some role in the granting of this subsidy. We know that Rutherford wrote the report on Moseley's request, but we also know that Paul Langevin had been invited to the ISC meeting. There is no doubt that Moseley's work had a strong influence on Rutherford's views on Bohr's atomic theory. This is testified by the following lines in Rutherford's communication to *Nature*, already cited in section I.5.3, ([Rutherford (1913b)], p. 423):

---

[10]See [Solvay XII (1962)], p. 16.

> The original suggestion of [Antonius] van der Broek that the charge on the nucleus is equal to the atomic number ... seems very promising. *This idea has already been used by Bohr in his theory of the constitution of atoms. Evidence in support of this hypothesis will be found in a paper by Moseley in the Philosophical Magazine of this month.* [emphasis added]

On December 3, 1913, i.e. three days before Rutherford's publication in *Nature*, a check of £ 39,59 was sent to Moseley by ISIP's Administrative Commission. The money could not have come at a better time. We quote Heilbron ([Heilbron (1966)], p. 352):

> Late in November, probably after writing up his results, Moseley moved from Manchester to Oxford to be with his widowed mother. Although he apparently did not have an appointment at the University, he was able to work in the physical laboratories, supporting himself by private means and his research by a subvention from the Solvay Institute.

Moseley thanked the Institute in [Moseley (1914)]. The paper was quoted by Arrhenius in a note recommending Moseley for the 1915 Nobel Prize in Chemistry. The Nobel Committee members were impressed, but argued that Moseley was young and that his Prize could wait. They did not foresee that this promising researcher would join the British army, and would die near Gallipoli on August 10, 1915.[11]

The Solvay archives contain an excerpt of Moseley's paper [Moseley (1914)], and a report dated April 3, 1914: Moseley indicates how he spent the money, underlining that the subsidy was used for the purchase of a *specially-designed* vacuum X-ray tube. He also gives a list of expenses incurred (including assistance and chemicals), and the reference of the paper that would be quoted by Arrhenius in his recommendation for a Nobel Prize.[12] We may therefore say that, without this equipment, Moseley would not have been able to finish a significant part of his work.

---

[11]See the following websites: `https://www.nist.gov/news-events/news/2024/02/henry-moseley-and-periodic-table-elementsand` `https://siarchives.si.edu/blog/science-service-close-two-haunting-portraits-henry-gwyn-jeffreys-moseley`.
[12]Archives de l'ESPCI, Fonds Langevin, doc.L12/83.

Henry G.J. Moseley (1887–1915) in his lab holding a glass globe
Courtesy of Smithsonian Institution Archives,
Acc. 90-105, Image no. SIA2009-1117

I.6.5 Last subsidies for the year 1913–1914

On February 2, 1914, the ISC proposed to grant three more subsidies:

- 1,000 francs to Laue for *the continuation of his research on the dispersion of Röntgen rays.*
- 1,250 francs to W. L. Bragg for *research on Röntgen rays.*
- 1,500 francs to Johannes **Stark** for *the continuation of his research on the splitting up of spectral lines in an electric field.*

The decisions on the Laue and Bragg subsidies do not require further comment. On the other hand, the support given to Stark deserves some words of explanation.

Stark

On November 20, 1913, Stark announced to the Prussian Academy of Science his success in observing the splitting up of spectral lines in an applied electric field [Stark (1913)]. Nearly two decades had passed since Pieter Zeeman's discovery of a similar effect in an applied magnetic field.

Galvanised by his finding, Stark rushed to Göttingen and to Berlin to inform his colleagues of his spectacular results. On his return trip he sent a letter to Lorentz,[13] explaining that he needed 20,000 marks to extend his measurements and increase their precision. Stark specified in the letter that, in the event of a favourable answer, he would personally thank Solvay for supporting his research. He even considered a one-day trip from Aachen to Brussels, to meet the industrialist and show him his spectrograms.

Lorentz, who new Stark since 1907, made him understand that it was up to the ISC to examine subsidy requests, and that he would not achieve anything by contacting Solvay. The advice had an unexpected effect: Stark replied that he would no longer ask for a grant, out of fear of overshadowing other requests. In a new letter to Stark ([Kox (2009)], p. 386), dated December 16, 1913, Lorentz explained that he would plea in his favour with the Committee members. The plea proved successful. Stark obtained 1,500 francs in February 1914 and was told that he would get an additional sum in June, taken from the budget year 1914-1915. The news materialised in the form of a cheque of 4,500 francs which was sent to Stark on June 24, 1914.

Comparing the amounts awarded by ISIP before the outbreak of WWI, we see that both Laue and Stark obtained a total of 6,000 francs. No other applicant received a similar amount. The size of the subsidies granted to the two German physicists reflects the extent of the interest their discoveries aroused among Lorentz and his colleagues. The events that followed proved them right: X-ray diffraction revolutionised the microscopic study of matter; Stark's discovery played a major role in Sommerfeld's extension of Bohr's atomic theory ([Kragh (2010)], p. 75).

Regarding Sommerfeld, we will see that he benefited twice from ISIP's subsidy programme. In 1913 he received 5,000 francs for enabling Friedrich to continue his research in Munich. In 1914–1915 he would benefit "indirectly" from the support given to Stark.

[13] Noord-Hollands Archief, archief H. A. Lorentz, 364, invent. num. 75.

I.6.6 Decisions taken by the ISC in June 1914

A meeting of the Scientific Committee took place on June 24, 1914. It was attended by Lorentz, Marie Curie, Brillouin, Nernst, Warburg and Knudsen. The chairman announced that after Stark's second subsidy, 20,500 francs were still available for the year 1914–1915.

The ISC examined the requests submitted before May 1914. One came from Wien, another from Wood. A third request was submitted by two Polish researchers: Jean Danysz, born Jan Kazimierz Danysz, and Ludwik Wertenstein. The Committee proposed to grant:

- 4,000 francs to Danysz and Wertenstein for *Research on radioactive recoil rays and on β-radiation by the radium family.*
- 4,000 francs to Wien for *Research on positive rays, and a spectral analysis of Röntgen rays.*
- 3,000 francs to Wood for *Further research on radiation.*

While Wien and Wood are familiar names, the Polish scientists Danysz and Wertenstein are less-known protagonists in the field of radioactivity.

Danysz had been praised by Rutherford at Solvay I for experiments that shed light on the complexity of the β-rays emitted by radioactive substances. During the discussion of Sommerfeld's report, Rutherford explained that Danysz had been able to observe "at least thirty distinct β-rays, emitted by a combination of two different varieties of radium B and C." ([Solvay I (1912)], p. 388). According to Carsten Jensen ([Jensen (2000)], p. 37):

> He [Danysz] was unknown in the scientific community in 1911 when he published the first results of his research; Rutherford simply referred to him as "a Frenchman" in a letter to Hahn in October 1911. By 1914, however, he was considered to be one of the leading β-spectrum researchers, and Rutherford and Stefan Meyer suggested him as the only participant, outside Berlin and Manchester, in the forthcoming congress on β, γ rays, and their origin.[14]

---

[14]The congress did not take place.

Danysz's growing popularity in the radioactivity field has been commented by his countryman Józef Rotblat ([Rotblat (1945)], p. 384). He underlined that Danysz and Wertenstein were considered the "two most outstanding pupils" of Marie Curie. Danysz was charged to prepare Madame Curie's courses during her illness (1911–1912).[15] In 1913, Danysz and Wertenstein defended their PhD thesis: the former focused on the β-radiation produced by the radium family, the latter on radioactive recoil. The same year, Madame Curie sent them both to Warsaw as personnel of a newly founded radiation laboratory. She accepted to be the *honorary director* and proposed Danysz as chief of staff. Continuing his research, Danysz showed, in collaboration with Wertenstein, that α-particles having penetrated the nucleus of a radioactive atom were powerless to alter the speed of radioactive transformations.

We quote Rotblat ([Rotblat (1945)], p. 384):

> In 1913, the Scientific Society of Warsaw founded, under the auspices of Madame Curie, the Radiological Laboratory, where research work on radioactivity was to be centralised. Madame Curie sent her two most outstanding pupils, Danysz and Wertenstein, to Warsaw to direct the work there. After the death of Danysz in 1914, Wertenstein became director of the Laboratory, and remained there until its destruction in 1939. ... *Despite the constant difficulties and limitations due to lack of funds*, he managed to create a centre of research which gained a high reputation abroad, and to found a school from which emerged a number of young people of high scientific standing. [emphasis added]

The Solvay subsidy clearly helped to compensate for the lack of funds. Danysz and Wertenstein received 4,000 francs on June 29, 1914, after having moved to Warsaw. They acknowledged receipt of the sum in a letter to Knudsen of July 2, specifying that the means would be used for the purchase of *a powerful electromagnet.*[16] Their aim was apparently to improve Danysz's magnetic deflection experiments,[17] but it is difficult to know how the magnet was used. We keep quoting Rotblat ([Rotblat (1945)], p. 384):

---

[15] See *Rapport de Madame Curie sur l'activité de son laboratoire pendant l'année scolaire 1912–1913*, Archives de l'ESPCI, Musée Curie, LC.MC-2450.
[16] Archives de l'ESPCI, Fonds Langevin, L010/245.
[17] See also [Jensen (2000)], p. 38.

> (In Warsaw) Jean Danysz was surprised by the war. He immediately took command of his section in the 28th infantry regiment. Having taken part in the Battle of the Marne, Danysz became flag-bearing officer of his regiment. On November 4, 1914, he fell gloriously at the head of his platoon near Cormicy, fatally wounded by a bullet in the chest.[18]

Going back to the ISC meeting of June 24, 1914, we end this section with the following comments:

(i) The granting of the three subsidies being decided, the chairman gave an account of the reports that had been received from previous Solvay grant recipients. Knudsen was asked to send a reminder to latecomers.
(ii) All by all, the Institute supported about 50 projects, during its first budget years, for a total amount of more than 80,000 francs.

As we will see in Part II, the Solvay Institute of Physics survived the collapse of international relations caused by the Great War. Unfortunately, the depreciation of the Belgian franc put an end to its subsidy programme. We may wonder what would have happened if, after WWI, the Institute had been able to continue to support the research that led to the birth of *Quantum Mechanics.*

---

[18] Extract from Livre d'or de l'École de physique et de chimie. See `http://vlecalvez.free.fr/Jean_Danysz.html`. Our tranlsation from the French.

PART II

Post-War Councils (1921–1927) and the Completion of Quantum Mechanics

Chapter II.1

Applications and extensions of Bohr's atomic theory

Peter Zeeman (1865–1943). Hydrogen's spectral lines splitting in a magnetic (Zeeman effect) and electric field (Stark effect) stimulated Bohr to improve his atomic model.
Courtesy of the Solvay Institutes

The Great War considerably slowed down research on quantum theory. However, discoveries that had been made during the years 1912–1914 led to major advances, in particular with regard to Bohr's atomic theory.

The assumption of atomic transitions with emission or absorption of a light quantum with a frequency determined by the difference in energy of the two levels, had an unexpected effect with far-reaching consequences. It inspired Einstein, who in 1916 recovered Planck's radiation law after introducing the concept of transition probabilities. Einstein also introduced the important idea of spontaneous emission.[1]

II.1.1 Impact of the Stark effect

Impressed by Stark's announcement of the splitting up of the spectral lines of hydrogen in an applied electric field, Rutherford sent this message to Bohr ([Bohr (1961)], p. 46): "I think it is rather up to you at the present time to write something on the Zeeman and electric effects, if it is possible to reconcile them with your theory."

Rutherford's words stimulated Bohr in his desire to account for a phenomenon that was not known when he published his theory. We quote Kragh ([Kragh (2010)], p. 71):

> Having acquainted himself with Stark's work, Bohr immediately took up the challenge and on the last day of 1913 he wrote to Rutherford "I have succeeded in accounting, at least partially, for the experiments of Stark on the basis of my theory". Bohr's paper, in which he introduced his theory of what quickly became known as the Stark effect, appeared in the March 1914 issue of the *Philosophical Magazine*" ... By the spring of 1914 Bohr was confident that his theory agreed with, or could be developed to agree with, the Stark effect.

In fact, an earlier attempt to explain the Stark effect in terms of the Bohr model had been made by Warburg, a member of the ISC. We keep quoting Kragh:

> Warburg, who mistakenly seems to have considered Bohr's theory as belonging to the tradition of the electromagnetic matter

[1]Enriched with the concept of *coherence*, Einstein's work would become a few years later the starting point of a theory that led to the *laser*.

theories, clearly found the theory appealing. But he was less happy about its foundation in Bohr's postulates, which invited "weighty misgivings". Although Warburg obtained a broadening of the right order of magnitude, he did not succeed in reproducing the distinct line patterns observed by Stark, from which he concluded that Bohr's theory was unable to give more than a partial explanation. Bohr disagreed. In a letter to Warburg of early 1914 he remarked that his own as yet unpublished calculations resulted in separate lines with the right separation in wavelength.

As pointed out by Kragh, Stark's discovery had also further effects:

> The work of Stark was influential in another case that soon became to be seen as a strong support of Bohr's theory, namely the Nobel Prize which rewarded the electron collision experiments of Franck and Hertz. That experiments of this kind might yield information about the energy levels of the stationary states of atoms was already suggested by Bohr in his trilogy, but the Franck and Hertz series of experiments were quite unrelated to Bohr's theory. Rather than thinking in terms of spectra and atomic structure, in agreement with an earlier idea of Stark, they interpreted their results as due to ionisation processes. In none of the publications of Franck and Hertz until 1916 did they refer to Bohr's theory, and when they did so in a paper of 1916 it was only to reject it.

In 1919, Franck and Hertz ended up acknowledging Bohr's theory. As specified by Kragh: "They received the 1925 Nobel Prize for Physics, in a large measure because of its connection with Bohr's atomic theory."

II.1.2 Sommerfeld's extension of the Bohr model

Bohr's most resounding success of 1914 was his demonstration that the lines of the Pickering–Fowler spectrum should be viewed as Balmer-like lines for ionised helium.[2] This feat made a strong impression on Einstein, who, upon hearing the news from George de Hevesy, declared ([Kragh (2010)], p. 19): "This is a tremendous achievement. Bohr's theory must then be right."

[2]The so-called Pickering–Fowler series consists of three lines of singly ionised helium found, usually in absorption, in the spectra of hot stars. The name comes from Edward Charles Pickering and Alfred Fowler.

Encouraged by the success, Bohr set out to extend his model. As noted by Eckert ([Eckert (2014)], p. 145), one of his objectives was "to account for minute corrections that would lead to a better agreement with Fowler's experimental results. In order to obtain such corrections he took into account elliptic orbits and a relativistic change of the electron mass".

Sommerfeld, the other major actor in this field, also thought of extending Bohr's model. His interest in the Bohr atom was indebted, at least in part, to the Stark effect and its relation to Bohr's theory.[3]

However, this interest was not exempt from criticism. In a letter of June 1914, Sommerfeld told Langevin ([Eckert (2014)], p. 143):

> Apparently, there is much truth in Bohr's model, and yet I believe that it has to be re-interpreted in a fundamental manner. At the moment, I find it particularly disturbing that it yields a wrong value for the magneton.

During the summer of 1914, Bohr travelled through Germany. In Munich he presented his theory on Sommerfeld's colloquium. It is likely that he spoke of the difficulties in explaining the details of the Stark effect, and that this admission reinforced Sommerfeld's quest for a "theory of the Zeeman and Stark effects". We quote Eckert ([Eckert (2014)], p. 151):

> Stark's experimental investigations on the splitting of spectral lines in hydrogen (Balmer lines) in electric fields, showed that the frequency of each Balmer line becomes decomposed into a multiplet of frequencies. The number of lines in such a decomposition grows with the index of the line in the Balmer series. Sommerfeld concluded from this observation that the quantisation in Bohr's model had to be altered in order to allow for such decomposition. He outlined this idea in a lecture in winter 1914–1915, but did not publish it. The First World War further delayed its elaboration. When Bohr published new results in autumn of 1915, Sommerfeld finally developed his theory in a provisional form in two memoirs which he presented in December 1915 and January 1916 to the Bavarian Academy of Science.

Eckert also underlines the fact that the discussions at Solvay I played an essential role in this achievement:

[3]See [Kragh (2010)], p. 75.

> The crucial step in Sommerfeld's work was not just to add ellipses to Bohr's circles and make use of relativity theory – both had been considered already by Bohr – but to invoke an approach that allowed the quantisation of several degrees of freedom. Sommerfeld resorted to a procedure that had been introduced by Planck at the Solvay Congress of 1911.
>
> On January 8, 1916, he presented a treatise to the Academy with far-reaching consequences. In this treatise he presented the relativistic theory of Kepler motion in which the orbit of an atomic electron is no longer closed, resulting in a fine-structure of spectral lines.[4] From the appendix to this treatise it is apparent that Sommerfeld had good reasons to assert his priority for the extension of Bohr's model. In this appendix he addressed papers by Max Planck, published in December 1915 on "The Quantum Hypothesis of Molecules with Several Degrees of Freedom", which used the same quantisation of phase integrals but felt short of Sommerfeld's results with regard to the detailed analysis of the spectra ...
>
> Planck was not primarily interested in Bohr's model or in spectral lines. His main effort addressed a deficiency in his black-body theory which operated with one-dimensional oscillators as sources of radiation. Henri Poincaré had alluded to this deficiency when he asked Planck at the first Solvay Congress in 1911, how one should quantise systems with several degrees of freedom

In December 1915, Sommerfeld succeeded in elaborating an extension of the Bohr model by quantising the eccentricity of ellipses in the same manner as the orbital motion. His main problem was the quantisation of atomic orbits that are deformed from the circular and elliptic shape. Using a hint from Karl Schwarzschild, he resorted to the Hamilton–Jacobi formalism and transformed the quantisation of phase integrals into a quantisation of action-angle variables ([Eckert (2014)], p. 151).[5] As noted by Pais ([Pais (1991)], p. 82), the astronomer Schwarzschild and Sommerfeld's student Paul Epstein showed independently in 1916 that the problem of the linear Stark effect is exactly solvable by means of the "Old Quantum Theory":

---

[4]In this paper Sommerfeld introduced his famous *fine-structure constant.*

[5]The Hamilton–Jacobi formalism's name comes from Sir William Rowan Hamilton and Carl Gustav Jacob Jacobi.

> The energy values of the split levels depend on more than one quantum number. The excellent agreement with experiment of their formula for those levels ranks next to the Balmer formula as the greatest quantitative success of the old quantum theory.

In 1918, the influence of Sommerfeld's work on Bohr became apparent with the first part of Bohr's comprehensive account "On the Quantum Theory of Line Spectra", in which his famous *correspondence principle* became the uniform point of view for the further developments of his ideas, with the hope of transforming his semi-classical model into a fully quantum one ([Eckert (2014)], p. 154).

Chapter II.2

The International Solvay Physics Institute after WWI

Maurice de Broglie (1875–1960). The X-ray work of Maurice de Broglie before Solvay III was an essential step in the development of quantum physics.
Courtesy of the Solvay Institutes

The Great War left deep marks on people's minds, particularly in Belgium and France. When in January 1919, the Institute was about to resume its activities, German physicists were no longer welcome in Brussels. This was partly due to the deception caused by the attitude of German physicists, who had been close to Solvay, or had handsomely benefited from ISIP's subsidy programme. Hence, the Institute's post-war situation required a change in its scientific management.

II.2.1 Temporary exclusion of German physicists

Ernest Solvay and the members of the Administrative Commission, were shocked to learn that Ostwald, Nernst, Planck and Wien had not abstained from signing the outrageous *Appeal to Civilised Nations* of October 1914, in which 93 German intellectuals declared their solidarity with the German army, denying the violation of Belgium's neutrality an the abuses committed by Germans soldiers in various parts of the country.

Lorentz, who from the start of the war expressed sympathy towards Solvay and Belgium, had been working all along to ease tensions between scientists from both camps. He intervened on several occasions with his German colleagues to soften the fate of Georges Hostelet, a Belgian prisoner of war, close to Solvay, and facilitate the writing of the Minutes of the second Physics Council. Aware of the difficulty of continuing his work within the Physics Institute, Lorentz sent a letter to Solvay,[1] to express his opinion on the attitude to adopt towards German physicists:

> I understand that at this moment the Belgian and French scientists no longer wish to have any relation with them. For the first years to come there will be no place for them in the Physics Institute. Those who attended the Councils will understand that they cannot come to Brussels ... However when speaking of Germans, we must not lose sight of the fact that there are representatives of all nuances. A man like Einstein, the great and profound physicist, is not "German" in the sense that the word is often used now. His judgment on the events of the past will hardly differ from yours or mine. There are also several physicists who refused to sign the scandalous Manifesto of the 93. In a conversation I had with him in the spring of 1915, Mr. Warburg strongly expressed his disapproval of the declaration.

[1] Letter from Lorentz to Solvay of January 10, 1919, Archives of the Solvay Institutes S.a.b.ULB, doc.683.

> Certainly, a number of signatories would still do the same if the opportunity presented itself, but I am sure that many others deeply regret the lightness with which they lent their name to the publication. You know the letter that Mr. Planck sent me, and which I published at his request ... All things considered, I believe I should suggest to you not to exclude all Germans, that is to say, not to close the door for ever. I hope that it can be opened to a new generation, and even that, perhaps, in the course of the years, we will be able to admit those scientists of our days of whom we can believe that they sincerely and honestly regret the events that took place. This is how German science will be able to resume the place which, despite everything, it deserves by its antecedents.

Solvay accepted the idea, as witnessed by a message from Tassel to Brillouin,[2] the French ISC member, known for his firm refusal to welcome German physicists at future Solvay meetings:

> Mr. Solvay is convinced that if personal relations between scientists belonging to belligerent countries will remain impossible for a long time, scientific relations can never be suppressed.

As we will see in sections II.2.2 and II.5.1, the difficulty of inviting German physicists[3] to the first post-war Physics Councils prevented ISIP's Scientific Committee from devoting Solvay III and IV to the development of quantum theory.[4]

II.2.2 Renewal of the International Scientific Committee

In his letter of January 10, 1919, Lorentz also addressed another delicate problem, ISIP's post-war reorganisation:

---

[2]Letter from Tassel to Brillouin of January 21, 1919, Archives of the Solvay Institutes, S.a.b.ULB, doc.675.

[3]An exception was made for Einstein, presented as a Swiss member. However, in 1921, he was not able to participate in Solvay II, and in 1924, as we shall see, he asked not to be invited.

[4]The exclusion of German scientists from the Solvay Councils and their rehabilitation process has been briefly addressed by Kenneth Bertrams in a paper entitled *Caught-up by Politics? The Solvay Councils on Physics and the Trials of Neutrality*. See [Lettevall *et al.* (2017)].

> How should the Physics Institute be reconstituted, and how should it get back to work? Of the nine members of the Scientific Committee, four should have left on November 1, 1916. The mandate of the other four would expire on November 1 of this year, while mine would last until 1922.
>
> Since the partial renewal, requested by the statutes for 1916, did not take place, and since we do not know which of the members should have left, the composition of the Committee does no longer conform to the statutes. I should therefore suggest to my colleagues on the Committee that we all offer our resignation to the Administrative Commission, or that we ask its members to consider us all resigners. For myself, it seems correct to me not to take into account my exceptional position. Together with you, the Administrative Commission will be able to choose a new Scientific Committee, which it will compose in the manner it deems appropriate on account of the current circumstances.

Solvay agreed. All mandates having expired, except that of the president, the ISC members were invited to present their collective resignation. No one was formally excluded. The two German members had no reason to feel humiliated. The renewed Scientific Committee remained under Lorentz's chairmanship. It comprised its former members (see section I.2.4), except Nernst, Warburg and Goldschmidt. The first two were replaced by W. H. Bragg and the Italian physicist Augusto Righi. As for Goldschmidt (the Belgian representative), he was invited to join the Scientific Committee of the renewed *International Solvay Institute of Chemistry*.[5] His seat in the Scientific Committee for physics was assigned to Edmond van Aubel.

Regarding the post-war recomposition of the ISC, it is worth emphasising that the Gauthier-Villars volume dedicated to Solvay II, which appeared in 1921, only provides information on the *renewed* Scientific Committee. Its pre-war composition does not appear in the first two volumes, a gap which adds to the lack of information on the subsidies granted on proposal of that Committee.

[5]This Institute had been founded in May 1913 in partnership with the International Association of Chemical Societies. It was only after the latter's dissolution that an International Scientific Committee for chemistry was created, in analogy with the one for physics.

Edmond van Aubel (1864–1941)
Courtesy of the Solvay Institutes

II.2.3 Planning the first post-war Council

The members of the new ISC met in Brussels on March 30 and 31, 1920. The discussions have been recorded in the "Report of the meetings of the Solvay Scientific Committee".[6] The chairman announced that, due to the increase in the costs of instruments and the depreciation of the Belgian franc, ISIP was forced to put an end to its subsidy programme. He told his colleagues that the French physicist Dunoyer, to whom a grant had been

[6]See the book "Comité Scientifique Solvay. Procès-verbaux", Archives de l'ESPCI, Fonds Langevin, L8/04.

awarded in November 1913, had been wounded at the beginning of the Great War and deprived from the possibility to receive a monochromator ordered in London. The manufacturer was willing to provide a new device, but at a much higher price. Lorentz explained that the Administrative Commission agreed to grant an additional 2,000 francs to Dunoyer.[7]

Despite ISIP's new financial situation, Tassel, proposed to continue supporting international projects "of exceptional interest". Reacting to Kamerlingh Onnes's idea of limiting the Institute's action to the support of national projects, he insisted on perpetuating Ernest Solvay's explicit wish. The Committee agreed to set up a financial reserve, to be used from time to time to support such projects. Tassel's comment is important because it again confirms that the establishment of the subsidy program was Solvay's idea.[8]

Regarding the organisation of future Physics Councils, Lorentz first wanted to settle two preliminary questions: "*Are these Councils really useful? Should we keep organising them?*" He declared that, in his opinion, the Solvay meetings had proved their efficiency in stimulating their members in their ongoing work, and in allowing them to embark on new paths of research. Lorentz mentioned, as a striking example, the decisive papers, written by Poincaré between December 1911 and the beginning of 1912, in reaction to a fundamental question that had been raised at the first Council, but had not been resolved.

Having obtained the agreement of his colleagues that ISIP would regard the organisation of Physics Councils as its main future task, the chairman insisted on the need to reduce the cost of these meetings and limit the number of the participants. Asked for his opinion, Rutherford underlined that "ordinary conferences are too big to be truly useful" and welcomed Lorentz's idea.[9] It was therefore decided that the number of invited members would be limited to twenty five, and that future Councils would be organised at a rate of one every three years.

Having agreed that Solvay III would take place in April 1921, the ISC members started discussing its programme. Lorentz announced that Bragg and Righi, who could not be present at the meeting, had made their opinions known by letter.

[7]See Lorentz's letter to Tassel (Secretary of the Administrative Commission) of August 18, 1919. Archives of the Solvay Institutes, S.a.b.ULB, doc.1968.

[8]See Verschaffelt's annotations of the meeting: doc. L11/52_bis ESPCI.

[9]See doc. L11/52_bis ESPCI.

- Righi wanted to address the experimental foundations of the theory of relativity. We know from a letter he sent to Solvay[10] on April 14, 1920, that his wish was to have new tests carried out in the hope that they would disprove Einstein's theory.[11]
- Bragg, on the other hand, explained that he was not in favour of discussing relativity. He proposed to focus on the structure of molecules, atoms and the atomic nucleus, and to address the origin of magnetism.
- Marie Curie, Rutherford, and Kamerlingh Onnes approved Bragg's idea, arguing that relativity theory did not raise pressing questions, while there were several problems in molecular theory "the discussion of which could lead to immediate and important progress".
- Marie Curie raised the question of the application of quantum theory, insisting on the presence of a "gulf" between theory and logic.
- Kamerlingh Onnes added that relativity was "too abstract a subject to be discussed at a Solvay meeting, and that magnetism was an issue that could be validly addressed in the absence of German physicists."

Lorentz closed the debate by proposing to devote Solvay III to the "role of electrons in various domains, ranging from radiation to the constitution of atoms and magnetism."

The ISC established a provisional list of 25 guests, representing eight countries: Lorentz, Kamerlingh Onnes, Wander de Haas, Zeeman, Ehrenfest (*the Netherlands*); Rutherford, W. H. Bragg, Owen Richardson, Jeans, J. J. Thomson, Barkla, Joseph Larmor (*United Kingdom*); Einstein (*Switzerland*); Marie Curie, Brillouin, Langevin, M. de Broglie, Perrin, Weiss (*France*); Bohr, Knudsen (*Denmark*); Lars Vegard or Manne Siegbahn (*Sweden*); Righi (*Italy*); Robert Millikan (*United States*); van Aubel (*Belgium*). In reserve: John Cunningham McLennan, Charles Fabry, Aimé Cotton, Wilhelmus Hendrikus Keesom, Léon Brillouin, W. L. Bragg, Samuel Jackson Barnett, Hantarō Nagaoka,[12] Émile Henriot.

Regarding Samuel Jackson Barnett, we should recall his discovery of the effect that bears his name, the production of magnetism in a rotating rod through the gyroscopic motions of the elementary magnets in the rod. He also made a very accurate measurement of the gyromagnetic ratio

[10] Archives of the Solvay Institutes, S.a.b.ULB, doc.752.
[11] In the following, we rephrased or translated ISC's "Procès-verbaux" – Archives de l'ESPCI, Fonds Langevin, L8/04.
[12] Nagaoka was the first physicist to develop a "planetary" atomic model. In 1900, he had taken part in the International Congress in Physics, organised in Paris.

for several materials – that is, the ratio of the magnetic moment of the elementary magnet to its mechanical momentum. At the time, existing theories predicted a value nearly twice as large as the one observed. This apparent anomaly led other physicists to the important hypothesis, and later to the proof, that the electron has spin momentum and a magnetic moment associated with this spin. By suggesting Barnett's name, Lorentz thought of a possible connection between his work, the electrons in the atoms, and the atomic spectra.[13] In the end, Barnett was not invited, but his work was mentioned by de Haas in his report. De Haas underlined that in his experimental work, Barnett described the reciprocal of the Einstein–de Haas effect ([Solvay III (1923)], p. 209).

Wander J. de Haas (1878–1960)

Owen W. Richardson (1879–1959)

Courtesy of the Solvay Institutes

The provisional agenda comprised the following reports:

1. *The classical electron theory*, by Lorentz.
2. (a) *Structure of the atom* (*charge and constitution of the nucleus, isotopes*), by Rutherford.
   (b) *Intervention of quantum theory* (*Einstein's relation in photoelectric phenomena*), by Maurice de Broglie.

[13]See Verschaffelt's annotations of the meeting: doc. L11/52_bis ESPCI.

3. *Bohr's spectral theory (distribution of electrons in the atom)*, by Bohr.
4. (a) *Electrons and magnetism (gyroscopic effect)*, by Einstein, Richardson[14] or de Haas.
   (b) *Attempts to explain para- and dia-magnetism; magnetism at very low temperatures; superconductivity*, by Kamerlingh Onnes or Langevin.

The Solvay Physiology Institute (on the left) and two other Solvay Institutes in Parc Léopold as pictured in a supplement to the journal *Le Temps* and published in *September 1920*. The building remained the stage of the Solvay meetings until 1927.
Courtesy of the Solvay Institutes

II.2.4 Maurice de Broglie, "rapporteur" at Solvay III

The programming of a report by Maurice de Broglie on X-ray photoelectric effects was, as we will see, an idea with far-reaching consequences. We already mentioned de Broglie's remark at Solvay II about the role that quanta should play in the interpretation of such effects. Let us now take a look at his work, and its impact on the research of his brother Louis.

Shortly after his doctoral thesis [M. de Broglie (1908)] on the ionisation of gases and the movement of particles suspended in gases, Maurice de

[14]It is worth remembering that Richardson was a recognised expert in electron theory and in the emission of electrons from hot bodies. In 1928 he would receive a Nobel Prize for his work (see [Kragh (2010)], p. 25).

Broglie turned towards an intensive study of X-rays.[15] Thanks to Laue's discovery of X-ray diffraction in crystals, a new and promising field of experimental research, with implications on the structure of matter, had opened up to physicists interested in the properties of this penetrating radiation. In 1913, Maurice managed to significantly improve the technique of X-ray spectrography. As noted by Jean-Jacques Trillat ([Trillat (1960)], p. 238):

> [Maurice de Broglie] imagined a method of extreme elegance, the rotating crystal method, thanks to which it became possible, by making a beam of X-rays fall on a crystal in slow rotation, to photographically record the spectra of diffraction, and thus obtain the spectral analysis of this radiation, and also essential information on the structure of the crystal.

From the beginning of his research, de Broglie discovered several fundamental laws that govern the emission and absorption of X-rays. He confirmed, in particular, Moseley's law by extending it to a large number of elements.

Soon after the end of the Great War, Louis de Broglie joined his elder brother at his Paris laboratory. In 1951, Maurice recalled the beginning of their collaboration ([Michel (1953)], p. 426):

> In the laboratory of Lord Byron street, ... he (Louis) found in the work I was carrying out on the X-ray photoelectric effect, a quantum phenomenon so curiously related to both waves and particles, a fertile breeding ground for his ideas.

Indeed, as recollected by Louis, the work with his brother Maurice laid the foundations of his discovery of the dual "wave-corpuscle" nature of matter ([Louis de Broglie (1931)], p. 16):

> Nor should it be thought that, by focusing on X-rays, I was loosing sight of my essential goal: the study of quanta. Indeed, the whole theory of high-frequency radiation is imbued with the quantum idea. Like light, these radiations behave as if they were made up of corpuscles of energy $h\nu$, the high value of the frequency meaning that the quantum $h\nu$ has a relatively large value, a fact

[15] In [M. de Broglie (1908)], de Broglie made use of X-rays to ionise the gases under study.

that facilitates its detection. On the other hand, X-rays originate in the deep regions of the atom, the architecture of which is essentially determined by the quantum of action, as we learned from Bohr's theory. Observation of the X-ray lines emitted by a given category of atoms provides information on the atom's internal structure, and on the movement of the electrons inside the atom, movements that are subject to quantum theory. Therefore, the study of X-ray spectra is closely linked to atomic theory and to the quantum mechanics of electron motion in the atom. So, it is easy to understand the interest it presented for me.

Chapter II.3

Solvay III (1921): Atoms and electrons

Solvay III participants and Ernest Solvay
Courtesy of the Solvay Institutes

The third Physics Council took place from April 1 to 6 at the Solvay Institute of Physiology. Lorentz welcomed the newcomers. They included Ehrenfest, Bragg junior, Wander de Haas (Lorentz's son in law), and the American Nobel laureate Albert A. Michelson.

Einstein was prevented from attending the meeting because of a trip to the United States. He nevertheless took part in the Council through de Haas, who presented Einstein's contribution as part of his report on electrons and magnetism. Bohr was exhausted by the creation of the Copenhagen Institute of Theoretical Physics. He decided not to come to Brussels, but arranged for his report to be presented by Ehrenfest.

II.3.1 Reports presented at Solvay III

The volume with the Minutes of Solvay III [Solvay III (1923)] mentions thirteen reports, many of which had not been scheduled[1]:

1. *On the electron theory*, by Lorentz.
2. *Structure of the atom* (*charge and constitution of the nucleus, isotopes*), by Rutherford.
3. *The relation $h\nu = E$ in photoelectric phenomena; production of light and Röntgen rays in collisions between atoms and electrons*, by Maurice de Broglie.
4. *Quantum absorption of radiation in metals*, by Robert Millikan.
5. *Paramagnetism at low temperatures considered from the point of view of the constitution of elementary magnets and the action they undergo from their carriers*, by Kamerlingh Onnes.
6. *The mutual action of magnetised molecules*, by Weiss.
7. *On superconductivity and the Rutherford–Bohr atomic model*, by H. Kamerlingh Onnes.
8. *Conductivity of metals*, by Léon Brillouin (son of M. Brillouin).
9. *The intensity of X-rays reflected by diamond*, by W. H. Bragg.
10. *The angular momentum in a magnetised body*, by de Haas.
11. *Application of quantum theory to atomic problems*, by Bohr.
12. *The correspondence principle*, by Paul Ehrenfest.
13. *On the arrangement and motion of electrons in atoms*, by Millikan.

[1]See the provisional list we presented in the previous chapter.

II.3.2 Highlights of the third Physics Council

The debates focused on two fundamental issues. One was about the then widely accepted Rutherford-Bohr model of the atom. The other concerned the dual "wave-particle" nature of light.

II.3.2.1 *The atom, "spinning" electrons, and isotopes*

In his report on the "Electron Theory", Lorentz considered the electron at rest as a sphere of radius R, carrying a charge e, uniformly distributed on its surface. Joseph Larmor, who this time accepted the invitation due to the topic of the Council,[2] insisted on his idea that electrons should be viewed as "singularities of the ether" ([Solvay III (1923)], p. 26), an idea shared by Lorentz.

Sir Joseph Larmor (1857–1942). Lucasian Professor of Mathematics from 1903 to 1932

Courtesy of the Solvay Institutes

[2]Lorentz and Larmor were the first to propose a theoretical framework to explain the concept of the electron. See [Arabatzis (2004)].

During the discussion that followed Lorentz's presentation, Rutherford made two remarkable statements. One was about the electron:

> Experiments carried out at the Cavendish Laboratory seem to indicate that the electron has properties other than its electric charge. Thus, for example, it seems to be endowed with a kind of polarity. The chief proof of this fact comes from the observation of the trajectories of β-particles obtained by Mr. C. T. R. Wilson's well-known method. This matter is treated in some detail in my report on the structure of the atom. In many cases these β-particles follow a spiral path, very different from what one would expect to find in random scattering. Compton suggested as an explanation that the electron has a magnetic polarity and that the reaction of the magnetic field induced in a slightly paramagnetic medium results in a spiral path. Although we are still at the stage of trying to explain the phenomenon, I am inclined to admit its reality.

Rutherford's other statement was about the observation of α-particles getting close to the nucleus in α-particle scattering by hydrogen atoms. Suspecting a failure of his "point-charge model" in collisions of close approach, he argued that forces, other than the electromagnetic, might be acting at the nuclear level: "I think that the inverse square law of distances is *completely at fault* at the distances existing between the positive and negative charges in the nucleus." This was one of the first hints that the nuclear forces should modify Coulomb's law when bullet particles approach the nucleus.[3]

As he reported on the *Structure of the Atom*, Rutherford reviewed all that had been learned since Solvay II, stressing the wealth of information acquired through the study of radioactive processes. He also recalled that in 1914 Darwin showed that Geiger and Marsden's results on the scattering of α-particles only agreed with the customary law of electric force, and that this law could be deduced directly from observations on the way in which scattering varied with a varying velocity of the incident particles.

In the following, we reproduce some of Rutherford's remarks at Solvay III from his original (unpublished) report in English.[4] They are organised in topical order.

[3]See [Feather (1977)].

[4]The report belongs to the set of archival documents kept at the Université Libre de Bruxelles, Archive des Instituts Solvay, S.a.b.ULB. A French translation of the quotes can be found in [Solvay III (1923)].

(i) *Regarding the atom and its nucleus*:

> We thus arrive at the important generalisation, strongly buttered by experimental evidence, that the main properties of an atom are defined by its atomic number which in turn is a measure of the resultant positive charge on the nucleus. The number and distribution of the outer electrons on which the main physical and chemical properties depend, is controlled by the nuclear charge, and only to a very subordinate extent by the mass of the atom. The latter is mainly concentrated in the nucleus and is dependent on the structure of the nucleus.
>
> As Moseley showed, the replacement of atomic weight by atomic number removes the outstanding difficulties in Mendeleev's periodic classification of the elements and throws additional light on the meaning of the periodic law. Very strong evidence of the correctness of this general point of view is obtained from the study of the radioactive elements. It seems clear that radioactivity is a property of the nucleus and that α and β particles, which play a fundamental part in the transformations, are expelled from the nucleus.
>
> The remarkably simple generalisation advanced by Russell, Fajans, and Soddy, shows that there is an intimate connection between the chemical properties of a product and the type of radiation from the parent element. The explanation of this relation is clear on the nuclear theory. The expulsion of an α particle (nucleus of a helium atom carrying two positive charges) lowers the nuclear charge or atomic number of the element by 2 unities and its mass by 4.
>
> The expulsion of a β particle from the nucleus raises the nuclear charge and the atomic number by 1 unit, but makes no appreciable difference in the mass. Knowing the type of radiation from the products and the atomic number of the head of the series, we can at once write down the atomic number and mass of all the elements in the series.
>
> These deductions are in close accord with the atomic numbers and atomic weights of these elements of the series which have been experimentally measured. This is a remarkable generalisation which has the same elements of simplicity shown by Moseley's relation between nuclear charge and atomic number. A simple

rule also connects the type of radiation with the position of the resulting element in Mendeleev's table.

(ii) *Regarding isotopes*

Sir Frederick Soddy (1877–1956)
Courtesy of the Nobel Foundation archive.

The idea of the existence of isotopes arose from a study of the radioactive bodies. A number of cases were early observed in a mixture of two substances, apparently impossible to separate by ordinary chemical methods, e.g. thorium and ionium, radium D and lead. Soddy made a systematic examination[5] of a commercial preparation of mesothorium, which contained radium, but found he was unable to effect a concentration of either component even after a large number of fractional crystallisations. The property of radioactivity was utilised to determine the relative proportions of

[5]See [Soddy (1914)].

radium and mesothorium before and after the chemical processes. He thus concluded that two elements might differ markedly in radioactive properties, but yet be so identical in ordinary physical and chemical properties that they are inseparable. He gave the name of isotopes to such inseparable elements, since they appeared to occupy the same place in the periodic table. The interpretation to be placed on isotopes was at once obvious on the nucleus theory. The isotopes must consist of atoms of the same resultant nuclear charge but with different internal constitution of the nucleus. Since the distribution and motion of the outer electrons, on which ordinary physical and chemical properties must depend, is almost entirely controlled by the charge on the nucleus, which for the distances involved may be regarded as a point charge, it followed that atoms might have different nuclear properties, e.g. mass and radioactivity, and yet be chemically identical, and even show almost identical spectra.

(iii) *On the β-particles scattering and a possible electron polarity.*[6]

On the nuclear theory, the laws of single scattering for the β-particle should be the same as for the α-particle, and consequently single scattering should predominate. Evidence is however accumulating that there is a marked distinction between the two cases. Anyone who has carefully examined the tracks of β-particles in Wilson's photographs cannot fail to have noticed an apparent regular curvature of all the tracks.

This is particularly obvious in the stereographic photographs of the tracks of β-particles liberated by the passage of an X-ray beam through air. As A. H. Compton has pointed out, a considerable fraction of these tracks appear to terminate in converging helices, which may either be right or left handed.

In an explanation of this observation, Compton supposes that the electron acts as a magnetic doublet whose axis remains approximately fixed in direction, and that the curvature of the path is due to the restriction of the magnetic field induced in the medium, supposed slightly paramagnetic ... The suggestion

[6]This point was already addressed by Rutherford during the discussion of Lorentz's report.

> that the electron has magnetic properties is a tentative one, but further and more convincing evidence will be required before such an important fact can be accepted. Assuming however, that the electron has such a property, it is natural to suppose that it is in some way due to spin of the electron.

II.3.2.2 *Rutherford's prediction of the neutron*

Towards the end if his presentation, Rutherford addressed the problem of the constitution of atomic nuclei:

> The study of radioactivity has learned that the nuclei of heavy radioactive atoms expel charged helium atoms and fast electrons. It is logical to conclude that the nuclei of atoms in general are formed partly of helium nuclei of charge 2 and negative electrons.

Rutherford's conclusion would hold several years, before being definitively abandoned after the discovery of the neutron. Regarding this discovery, made in 1932 by his co-worker James Chadwick, it is interesting to note that eleven years earlier, during the discussion of his Solvay report, Rutherford suggested the existence of an electrically neutral particle,[7] that he called a "neutron". Reflecting at the time on the composition of gaseous nebulae, he made this premonitory remark:

> There is, in fact, an enormous quantity of energy released by the combination of hydrogen nuclei to form helium ... The idea came to me that the hydrogen in nebulae comes from particles that we could call "neutrons" and which are formed of a positive nucleus of hydrogen with an electron at a very small distance. These neutrons would exert little action upon entering matter. They would serve as intermediaries in assembling the nucleus of high-atomic weight elements.

Reacting to Rutherford's prediction, Maurice de Broglie made this quite pertinent comment: "Particles with zero total charge, such as Sir Ernest's neutrons, easily escape our means of investigation; they could play a role in explaining the still unknown causes which determine the dislocation of nuclei."

[7] At the time, there was no indication that an electrically neutral particle could exist.

In fact, at Solvay III, Rutherford underlined what he had already predicted in his Bakerian Lecture of 1920 at the Royal Society, namely the existence of an electrically neutral particle in order to account for the existence of isotopes. Shortly after Chadwick's discovery of the neutron, Frederic Joliot told Arthur Stewart Eve ([Eve (1939)], p. 359) that although he and Irène Curie followed all publications "with care and interest", they had not read Rutherford's Bakerian Lecture. Joliot stated that if he and his wife had read Rutherford's prophetic suggestion about the neutron, it is possible and probable that they would have identified it. But Rutherford's prediction also appeared (in French) in the Minutes of Solvay III, published in 1923 by Gauthier-Villars. We therefore conclude that the little attention paid to these Minutes – a point emphasised and deplored by Pauli[8] – may have deprived the Joliot-Curies of the discovery of the neutron.

II.3.2.3 *Beginnings of the "Heisenberg line" of research*

The Rutherford-Bohr atomic theory played also a fundamental role in another context. It was the starting point of a process, developed by Heisenberg, Born and Jordan, that led to a "quantum mechanics" in terms of matrices. As pointed out by Heilbron [Heilbron and Rovelli (2023)], this process can be traced back to Hendrik A. Kramers's doctoral thesis of 1919. In his dissertation, entitled "Intensities of spectral lines", Kramers used Bohr's correspondence principle to connect the intensity of atomic spectral lines with the amplitudes of the harmonic vibrations representing the atom's electric moment. These amplitudes would later become the Heisenberg– Born–Jordan matrix elements. At Solvay III, Langevin suggested considering the atom as an infinite "ensemble of resonators", because of its ability to emit an infinite sequence of frequencies. During the discussion of Bohr's report (presented by Ehrenfest), Langevin explicitly referred to the Kramers thesis.

II.3.3 Maurice de Broglie's verification of Bohr's theory

At the beginning of his Solvay report, Maurice de Broglie made this striking remark[9]:

---

[8]See section Foreword, p. xi.

[9]We restart translating in English the French proceedings of Solvay III.

> Photoelectric phenomena are among the first interpreted by means of quantum theory. The emission, by an illuminated body, of corpuscles whose energy depends only on the frequency of the exciting radiation, and the very remarkable fact that the electrons emitted by a body under the influence of X-rays have an energy of the same order as that of the cathode electrons in the tube generating the radiation, immediately provides material support for the Planck–Einstein theory.

In fact, one of the problems discussed at Solvay III was whether an atomic electron could absorb an entire quanta of incident radiant energy. This problem was solved during the meeting, as indicated by the following exchange:

> **Lorentz:** The main question is this: Do the electrons that leave the atom have a full quantum of energy or not?
>
> **Maurice de Broglie:** My latest experiments are quite clear from this point of view: there is reason to subtract an output of work from the quantum of the electrons ... In short, the phenomena where it is necessary to subtract an output of work corresponding to the Bohr levels form a very clear whole which seems to constitute the main part of what is happening.

Lorentz and Maurice de Broglie sitting on the far left and the far right in the official picture of Solvay III
Courtesy of the Solvay Institutes

De Broglie's statement was the result of several months of intense research. We quote Wheaton ([Wheaton (1983)], p. 266):

> Maurice de Broglie was skeptical of the claim that the X-ray energy absorbed by electrons is always an integral number of $h\nu$. When he first took up the detailed evaluation of electron energies in February 1921, he thought he had found evidence that electrons leave the atom with quanta of energy $h\nu$, and therefore must have absorbed more than that quantised amount from the X-rays. His study of X-ray absorption edges and his understanding of the Bohr atom made it clear that some energy must be spent to remove the electron from the atom. This seemed a significant issue to clarify for presentation at the Third Solvay Congress, to which he had been invited and which would meet in three months of time ...
>
> He quickly adapted and modified the Manchester β-ray velocity spectrometer and employed it to analyse X-ray absorption, using what he called "particle velocity spectra" ... In collaboration with his younger brother Louis, Maurice de Broglie immediately formulated the goals of his research in terms of the Bohr model. By April 1921, they agreed with Simons and Richardson that Barkla was wrong.[10] Each electron in an atom, granted energy $h\nu$ by the incident monochromatic X-ray beam, must spend a certain amount of that energy to escape the atom. The amount was always just the "depth" of the orbit from which the electron came, and those depths had been carefully charted in de Broglie's private laboratory in rue Chateaubriand ... Measurement of the kinetic energy of the released electrons therefore yields directly the energy levels of the vacated Bohr orbit. In this sense, the X-ray photo effect could not be independently verified, but the systematic deviations from the quantum relation could be explained by an appeal to the growing understanding of the electronic structure of the atom ... De Broglie began these studies in order to shed light on the atom and to verify his previous X-ray absorption results.

[10]Lewis Simons was a South African physicist, who, like Richardson was convinced that Barkla's anti-quantum claims were simply wrong. M. de Broglie knew of Simons' paper of 1921.

II.3.4 Photo-effects and the light quantum

We keep quoting Wheaton[11]:

> Before 1921 the clearest evidence for spatial localisation in radiation came from Millikan's results for the photoelectric effect. But this case of quantum absorption concerned visible light, in which the localised properties were not strongly pronounced and the resistance of orthodox wave theory was stiffest. Furthermore, Millikan's data concerned only one point in the spectrum of photoelectric velocities: the maximum velocity.
>
> In this period, Millikan himself opposed the light quantum hypothesis, calling it "reckless","untenable"and even "erroneous". Despite his uncontested proof of Einstein's linear photo-effect equation, Millikan never doubted, until 1922, that an explanation of the result would be found without the suspect light quantum ... By April 1921, when he reported his results to the Third Solvay Congress, de Broglie was also aware of their significance for the photoelectric relations. Millikan had been able to compare optical frequencies to the maximum electron velocity; de Broglie could offer corroboration at points within the electron distribution as well. Thus, introducing corpuscular spectra as a means to analyse the atom, he proceeded to a general discussion of the photoelectric effect.
>
> Assuming the validity of the Bohr-Kossel model for X-ray emission, it was clear that the full energy quantum of the X-ray is transferred to individual electrons.[12] Coupled with the evidence that each X-ray frequency gives rise to a particular set of electron velocities, it became virtually impossible for de Broglie to avoid the issue of the light quantum. On several occasions in his presentation he alluded to the "remarkable" behaviour of the absorption process when the energy of secondary electrons is found to be the same as that of the entire spherical X-ray wave.

---

[11] See [Wheaton (1983)], p. 258 and 269.

[12] Walther Kossel had extended Moseley's work on X-ray spectra.

Robert Andrews Millikan (1868–1953)
Courtesy of the Solvay Institutes

II.3.5 Towards a recognition of the dual nature of light

Two decades after Planck's discovery of the quantum of action, there was a feeling that this universal constant, which had been shown to play a fundamental role in the constitution of atoms, should also be responsible for the complex nature of light. This view, first introduced with Einstein's light quantum hypothesis in 1905, and reinforced in 1909 by his "fluctuation formulas", was defended at Solvay III by Richardson and Bragg. However, the strongest arguments in favour of a dual nature of light were put forward by Maurice de Broglie. We still quote Wheaton ([Wheaton (1983)], p. 263):

> It is not entirely surprising that the earliest reaffirmation of the dualism inherent to radiation came neither from Britain nor Germany. There the issues had been recognised, discussed, and dismissed over the preceding twenty years. And American research traditions were largely derivative of German and British models: [William] Duane and Millikan had both studied under Planck, and

> Richardson did research with Thomson in Cambridge. It took an outsider to these traditions to raise the issues in a way that began to convince the others. That process began in France.

Indeed, the major change in the perception of the duality of light had been brought about by the study of X-ray absorption and the development of β-ray spectroscopy. We quote Kuhn[13]:

> With these new data in hand, it became possible to strike precise balances between the energy lost at the target by the cathode-ray electron in an X-ray tube, the orbital energies in the interior electrons of the target atom, and the energy $h\nu$ (and thus the frequency v) of the resulting X-ray. In the laboratories where these techniques were developed and these energy balances observed it was impossible not to notice that, with respect to spatial localisation and energetic relations, radiation behaved precisely as particles. Those conditions were not, however, fulfilled before 1920, and then only in a few places. One of these was the private laboratory set up in his Paris home by the French nobleman, Maurice de Broglie. By 1921, de Broglie's laboratory had mapped the X-ray energy levels of many atoms, and during that year an improved β-ray spectrometer was used to study the velocities of the photon-electrons ejected by X-rays of known frequency. Reporting the results to the third Solvay Congress in April of that year, de Broglie insisted that the radiation must be corpuscular, or, if it is undulatory, its energy must be concentrated in points of the surface of the wave.

In fact, defending the idea of "needle shaped X-rays", de Broglie discussed a number of phenomena in which quanta seemed to intervene as "individual entities". In this sense, his analysis of photo-effects and inverse processes, brought an answer to the remark made by Marcel Brillouin at the end of Solvay I (see section I.1.5):

> It isn't very satisfactory to be reduced to recognising the presence of a discontinuity by means of apparently continuous phenomena, to introduce it at the basis of a theory, before drowning it out with the help of statistical considerations... If we could imagine an experiment which made us seize the discontinuity "on the spot," it would be much more decisive and instructive.

[13] See Thomas Kuhn's preface to [Wheaton (1983)], pp. x–xi.

Particularly striking was de Broglie's conclusion from his X-ray results:

> An atom which receives light of frequency ν emits a projectile, endowed with energy $h\nu$, long before the radiation has been able to provide it with elements of this energy by means of homogeneous spherical waves.

II.3.6 Legacies of Solvay III

II.3.6.1 *Einstein's and Bohr's Nobel Prizes (1921, 1922)*

By validating Einstein's expression for the energy of a light quantum, and by carefully checking the energy balance in the release of atomic electrons, Maurice de Broglie provided decisive support to Bohr's theory. His Solvay report made a great impression on Manne Siegbahn, the first Swedish invitee to a Physics Council. Back in his country, this expert on X-ray spectroscopy shared his enthusiasm with Carl Oseen, a friend of Bohr and a fervent Einstein supporter. Elected member of the Nobel Physics Committee in 1922, Oseen became the architect of two Nobel Prizes: the one of 1921, awarded to Einstein "for having established a link between quanta and the photoelectric effect", and that of 1922, awarded to Bohr "for having discovered the link between quanta and the structure the atom". Details on the awarding of these Prizes can be found in Wheaton's book. They illustrate Oseen's strategy, and the impact of de Broglie's Solvay report ([Wheaton (1983)], p. 280):

> In his letter of nomination (for Einstein) Oseen quoted with emphasis Manne Siegbahn's interpretation of Einstein's law that energy transformation from radiation to matter, or vice-versa, occurs only in multiples of Planck's constant times frequency. Siegbahn's opinion was of particular significance because of his knowledge of, and research on, X-ray absorption features, essential preparation for interpretation of the X-ray photo-effect.
>
> Much of Einstein's theoretical work did not satisfy the criteria established in the Nobel will for an accomplishment of practical significance. But the theoretical underpinning of the photocell, already in industrial use in the post-war period, was eligible. It was no coincidence that Einstein's award was the first Nobel Prize given after de Broglie's and Ellis's detailed verification of the photo-effect

law for radiations other than visible light.[14] The Prize normally to be awarded in late 1921 was reserved and then given to Einstein in late 1922, along with the 1922 Nobel Prize to Niels Bohr.

In his precise of Einstein's work for the 1921 prize, Oseen lauded the "original and perceptive analysis" that led to Einstein's claim that an atom can absorb only multiples of single quanta. This hypothesis, he said, was given the"most beautiful confirmation" by Millikan's experiments on the energies of the photo-electrons, but there was no explicit appeal to de Broglie X-ray work in the justification for the Einstein award. That was reserved for the place where it would have maximum impact: the brief for Bohr's award written by Oseen at the same time.

Oseen saw the two prizes as inextricably linked: in supporting Einstein's, he was furthering Bohr's. De Broglie's X-ray evidence formed the crucial empirical bridge between the two. The Einstein-Bohr $h\nu$ law ... linked Bohr's atomic theory to the photo-effect. If the law of the photo-effect was valid, de Broglie's results verified Bohr's atomic structure down to the innermost shells of the atom. Thus, Oseen first used Millikan's confirmation of the photo-effect law to justify Einstein's award, and then appealed to the de Broglie's results in justifying the award to Bohr. The Royal Swedish Academy agreed, although only after holding off the Einstein award until both could be granted in 1922.

The award of the 1923 Nobel Prize to Millikan for his precise experimental work, including his verification of the photo-effect law, was entirely consistent with the position taken by the Academy. The awards of Einstein and Bohr gave implicit sanction to the light quantum hypothesis of radiation absorption that underlay the photo-effect law.

II.3.6.2 *Impact of Maurice's work on Louis's research*

We expanded in section II.2.4 on the collaboration of Maurice and Louis de Broglie. In this section, we want to say a few words on the impact of Maurice's work on his brother. We start quoting Wheaton ([Wheaton (1983)], p. 275):

---

[14]Charles Ellis was a Rutherford student at Cambridge. At Solvay III, Rutherford announced that Ellis's results for γ-rays were in full agreement with those of de Broglie ([Wheaton (1983)], p. 271).

Although the younger de Broglie's interests in 1920 centered on atomic theory, within a year they had changed significantly. Maurice's experiments showed with increasing clarity in 1921 that an X-ray transfers its entire quantum of energy to individual electrons, and Louis began trying to find an explanation. Years later he recalled the influence that his brother's research had had on him: "It directed my attention to the quality of waves and of X-rays and γ-rays." He told his brother:

> We debated the most pressing and baffling questions of the time, in particular the interpretation of your experiments on X-ray photo effect ... The insistence with which you directed my attention to the importance and the undeniable accuracy of the dual particle and wave properties of radiation little by little redirected my thoughts.

In 1921, following his brother's report to the Solvay Congress, Louis turned his attention to the counter-intuitive behaviour of X-rays, that, in the words of his brother [Maurice], concentrates energy "in points of the surface of the wave".

A little known incident, reported by Maurice de Broglie, sheds light on the situation at the third Physics Council, and on his brother's desire to learn more about the developments in quantum theory. Having taken an active part in Maurice's investigations, Louis expected to be allowed to attend the meeting as an auditor, a request turned down by Lorentz. Here are Maurice's words, translated into English ([M. de Broglie (1951)], p. 43):

> We were then in one of these periods where the balance between theory and experiment, the ordinary source of progress in the physical sciences, favoured experiment, and one clearly feels when re-reading Lorentz's report on electron theory, and the discussion that followed, how much the theoretical ideas were then overwhelmed by the facts. It was customary in Brussels to allow a few people from outside the Council, but versed in physics, to attend the sessions.[15] A young man, at that time almost unknown, requested this authorisation which was refused. He keenly felt

[15] Lorentz firmly refused to admit auditors. The only "outsiders" to which de Broglie refers were Solvay's close collaborators (Édouard Herzen and George Hostelet).

disappointed by this failure and vowed to return to the next Council, but this time through the front door. It was Louis de Broglie, who a few years later was to hold one of the first places in the Brussels meetings, by exposing the new quantum dynamics that would restore the pace of theoretical conceptions rejuvenated by his work.

During the discussion of de Broglie's report, Millikan presented a note "On the absorption of radiation by quanta". Referring to the fact that absorption and emission of radiation occurred by quanta, but without assuming the quantised nature of light itself, he defended the view that metals contain free electrons that can be put into motion through absorption of radiant energy, with a velocity determined by the quantum rule. Reacting to Millikan's insistence on the intrinsic capability of the electron to absorb and emit quanta of radiant energy, Maurice de Broglie made the following statement:

> Whatever the point of view, it seems difficult to admit that a free electron can be set into motion by radiation according to the quantum rule, unless one accepts the idea that radiation itself has a structure quite different from that represented by the two classical vectors (of classical electromagnetic theory).

In a paper on the photoelectric effect published a few months after the Solvay meeting [M. de Broglie (1921a)], de Broglie introduced an "electron frequency" by equating two quantities (in the spirit of Einstein's photoelectric law): a quantum of radiant energy, determined by the X-ray frequency, and the kinetic energy of the expelled electrons.[16] His viewpoint was based on the research completed for Solvay III on X-ray corpuscular spectra, which provide a "kinetic transposition of the spectrum belonging to periodic vibrations" [M. de Broglie (1921b)], p. 1158). He further presented his results and ideas in a book published in 1922.[17] In section 2 of this book, devoted to "periodic radiations", de Broglie made the following remark[18]:

[16]This frequency resembled the intrinsic electron property mentioned by Millikan.
[17]See [M. de Broglie (1922)], p. 20.
[18]We quote the English translation of the book which was published two years later. The only addition, which we do not quote, concerns the Compton effect.

> Certain facts even lead us to think that the vibrational energy, instead of being uniformly distributed over a wave surface, is condensed at certain points so that, there also, we are very close to the localisation of the energy which characterises a moving projectile. Adding to this the existence of inertia for energy, we see how closely we approach the emission theory (of light).

In section 3 of his book devoted to "corpuscular radiation" [M. de Broglie (1922)], Maurice declared:

> The study of rays from radioactive substances, and the analysis of electrical discharge through rarefied gases, have led to the discovery of radiations of a completely different nature than the periodic radiations known until then. In particular cathode rays and β-rays are made up of moving electrons, for which we can also define a quantum equal to their kinetic energy. In section 2, we defined a quantum for periodic vibrations, representing an energy independent of the intensity of the rays; now we assign to corpuscles with a definite kinetic energy, a quantum equal to this energy. This quantum divided by Planck's constant is a frequency ... The consideration of the quantum introduces a sort of link between wave radiations and corpuscular radiations; it also expresses what remains constant when, in contact with matter, these radiations are transformed into one another.

Maurice de Broglie's idea to associate a "frequency" to the kinetic energy of an electron was reported by the historian Olivier Darrigol. He noticed that Maurice believed that "both electrons and X-rays possess wave and corpuscular characteristics," an idea triggered by the "extremely detailed correspondence between [electron's] corpuscular spectrum and X-ray spectra" ([Darrigol (1993)], p. 321). The point we wish to emphasise is that this association was the fruit of research carried out before Solvay III, and further investigations, stimulated by the discussions that took place at the Physics Council. In [Rocci and Lambert (2024)], we investigated in more detail the connection between Maurice's work and the evolution of his brother's ideas. Here, we just say that the undulatory nature of corpuscular spectra was considered a consequence of the interaction with electromagnetic radiation and not an intrinsic characteristic of the electrons, as in Louis's future work ([Rocci and Lambert (2024)], p. 19).

II.3.6.3 *On "localisation" of X-ray energy*

Inspired by the results of his analysis of electrical discharges in rarefied gases, Maurice de Broglie concluded his book on X-rays with the following remarks ([M. de Broglie (1922)], p. 185):

> A spherical X-ray wave, expanding to increase its radius within a gas, undergoes a loss of energy through absorption, and we observe this:
>
> 1°: Among the molecules encountered, only a very small number react to undergo ionisation and expel a fast electron.
>
> 2°: The energy of the electron thus expelled by an atom is considerably greater than what it would be able to receive from the wave, by collecting the elements carried by the portion of the wave that passes in its neighbourhood.
>
> We are therefore led to think that, in order to be ionised, the atom must be in a particular "receptive" state, or that the energy of the incident wave is not distributed uniformly over its spherical surface. The way in which the energy of frequency ν is captured by the atom remains totally mysterious; we do not understand any better, in the case where the corpuscle's energy comes from the internal energy of the atom, how it would present such a close relationship with the frequency of the incident wave, since the wave would then only play a triggering role.
>
> The same considerations can be applied to the photoelectric effect of X-rays with respect to solid bodies. The fundamental point is always the fact that the velocity of the expelled electrons depends on the frequency of the exciting radiation, and not on its intensity. If we were to admit, as has sometimes been proposed (Sir Joseph Thomson), the localisation of energy at certain points on the wave surface, we would be very near to a corpuscular theory of light.

The last sentence seems to indicate that shortly after Solvay III, Maurice de Broglie had gathered compelling evidence in support of a dual nature of any type of radiation, periodic or corpuscular. However, he refrained from taking the final step. That step would be taken a few months later, under Maurice de Broglie's influence, by two younger researchers, born in 1892: the American physicist Arthur Holly Compton, and Maurice's theoretically minded brother Louis.

Chapter II.4

Origin of the Compton effect

Arthur Holly Compton pictured with the central result of his quantum scattering theory, as written in archaic version notation

Courtesy of the Washington University Photographic Services Collection, *Julian Edison Department of Special Collections*, Washington University Libraries

It is interesting to note that as the third Physics Council came to a close (April 6, 1921), Arthur Holly Compton, an experimental X-ray researcher at Washington University, was about to discover an effect that would prove crucial to the quantum revolution. Rutherford mentioned his name when discussing the electron "polarity" suggested by the spiral trajectories of β-particles. Let us now see how Compton was led to his discovery.

II.4.1 Compton and Barkla's results on secondary X-rays

Compton was guided in his research by earlier results obtained by Barkla. As mentioned at the end of section I.3.2, he studied a phenomenon that had been at the centre of Barkla's concerns: the "scattered rays" that are part of the secondary X-radiation produced by the impact of primary X-rays on matter. The role of Barkla's work has been recounted by Manne Siegbahn in his "Presentation Speech" at the Nobel Award Ceremony of December 10, 1927. These are his words[1]:

> Soon after Röntgen's discovery, it became known that matter exposed to X-rays emits radiations of different character. Besides an emission of electrons, corresponding to the photo-electric effect, known also in the optical region of radiation, there is also a secondary X-radiation. Even before the methods of X-ray spectroscopy were known, these secondary X-rays were proved, by the investigation of their absorption, to be of a twofold nature. It was Barkla who, through his fundamental researches, proved that the secondary X-radiation consists partly in scattering of X-rays (diffused X-rays), which he thought to have the same penetrability as the original radiation, and partly in a specific radiation which was characteristic of the chemical atom, and which was more easily absorbed. When X-rays fell upon matter with small atomic weight, as for example graphite, Barkla was not able to detect the mentioned characteristic X-radiation, but only scattering; and consequently the secondary rays ought to have the same properties as the original X-rays. Barkla, however, had already been able to show that in this case also the secondary X-rays – at least partially – are more easily absorbed than the original radiation, and therefore have a greater wavelength.

[1]See Appendix D for an extended version of the quotation.

Barkla thought this to be a new characteristic X-radiation. This is the point where Compton came in ... He made exact spectrometric investigations of the secondary X-radiation from matter with small atomic weight: in other words, he undertook to investigate exactly the scattered X-radiation. After some preliminary work, he found an experimental method that gave results as exact as they were astonishing. Using homogeneous X-rays corresponding optically to monochromatic illumination, that is to say, using a source of light that emits only one single spectral line, he found that the scattered radiation consists of two lines, one exactly the same as that of the source of rays, the other with a somewhat greater wavelength. This was the first evident manifestation of the Compton effect.

Manne Siegbahn (1886–1978)
Courtesy of the Solvay Institutes

II.4.2 Compton and the de Broglie brothers

Compton was influenced in his investigations by Maurice and Louis de Broglie. In October 1922, he wrote a report on the "secondary" radiations produced by X-rays [Compton (1922)], in which he referred to earlier works of Maurice and Louis (he also mentioned receiving a letter from Louis). By "secondary X-rays", Compton meant any radiation of the X-type excited by the passage through matter of primary X-rays. He pointed at the difficulty to distinguish experimentally, in certain cases, between two types of secondary radiation: scattered radiation, i.e. radiation emitted by the electrons in matter due to accelerations to which they are directly submitted by the primary rays, and fluorescent radiation, i.e. radiation of the energy absorbed from the primary beam and stored temporarily in the kinetic and potential energies of the electrons.

In part I of his paper, devoted to general fluorescent X-rays, Compton came to the following conclusion:

> Thus we see that the classical electrodynamics succeeds in explaining quantitatively many of the phenomena of secondary X-radiation, supposing that a considerable part of this radiation is truly scattered. The change in wave-length of the rays as they are transformed from primary to secondary X-rays seems to be in accord rather with quantum principles.[2] But it has not been possible to account on either basis for all the observed phenomena.

Compton's idea of attributing a quantum character to the shift in frequency, observed in the scattered X-rays, was based on Maurice de Broglie's report at Solvay III, even if he did not mention it. Indeed, as Compton applied quantum theory to what he called the Doppler effect in secondary X-rays, he assumed "that each electron, when it scatters X-rays, receives a whole quantum of energy and radiates back the whole quantum in a definite direction".[3] Now, as we have seen in the previous chapter, the transmission of an "entire" quantum in the interaction between radiation and electrons was a central point in de Broglie's Solvay report.

---

[2] According to classical electron theory, scattered X-rays should be of the same wavelength as primary X-rays.

[3] The Doppler effect is named after the Austrian physicist Christian Doppler.

In part II of his paper, devoted to the methods of experimental investigation of the photo-electrons produced by X-rays, Compton highlighted the benefits offered by a Wilson cloud chamber:

> In many respects the most satisfactory method of studying these secondary electronic rays is the beautiful one employed by C. T. R. Wilson, in which the tracks of the individual particles are rendered visible by condensing water droplets on the ions formed along their paths. By this means it is possible to count accurately the number of secondary electrons emitted, study their distribution, and make measurements of their range in air ... For investigating the velocities of the photo-electrons excited by X-rays, the method of photographing their magnetic spectrum has given the best results ... A very good description of the details of this method is given by de Broglie[4] in his paper on "The Corpuscular Spectra of the Elements" ... Finally a series of beautiful experiments by de Broglie in the X-ray region[5] and by Ellis in the region of γ-rays, both using the magnetic spectrum method, has shown that at least a large part of the electrons emitted from different energy levels within the atom absorb one quantum of the incident energy, and emerge with their kinetic energy diminished only by the work required to leave the atom.

Finally, in part III of his paper, devoted to the absorption of X-rays, Compton stressed again the difficulty of accounting for certain facts by means of quantum principles:

> The absorption of high frequency radiation is due at least to two independent processes. The more important of these is usually the energy spent in exciting photo-electrons, and resulting in fluorescent radiation ... This part of the absorption represents the energy which, as we have seen, is transformed into kinetic energy of the photo-electrons in accord with the quantum relation ... These facts, especially the reappearance of the absorbed energy in quanta of kinetic energy, would suggest that X-ray absorption is essentially a quantum phenomenon, to which presumably the

[4]See [M. de Broglie (1921a)], p. 265.
[5]See [M. de Broglie (1921a)], pp. 274, 746 and 806.

> classical electrodynamics may not be successfully applied. It is therefore surprising to find that quantum principles have offered no suggestion as to the significance of Owen's simple law,[6] whereas the classical electrical theory may be applied to the problem with striking results.

Then, after presenting his own derivation of a relation equivalent to Owen's law, Compton concluded:

> The fact that without introducing the quantum concept we can thus obtain a formula which is in many ways so satisfactory, strongly suggests that X-ray absorption is not a quantum phenomenon. The results rather support the view that X-ray absorption is a continuous process, obeying the usual laws of electrodynamics.

He nevertheless added:

> Since the above discussion was written, there has appeared a theory of X-ray absorption by L. de Broglie [L. de Broglie (1920, 1921, 1922)], which, although based upon widely different assumptions leads to nearly the same result ... It is certainly remarkable that results so nearly identical should be obtained on the basis of wholly different assumptions.

In a subsequent report, Compton examined a third type of absorption, prominent at very short wave-lengths. He presented an interpretation of the "true absorption" of γ-rays, based on the idea that when an electron receives a quantum of γ-ray energy, the "*momentum of the quantum*" gives a considerable momentum to the scattering electron, so that the scattered energy is equal to the energy absorbed less the kinetic energy of the recoiling electron. Compton called this phenomenon "*momentum absorption*", since it seemed to be due to the momentum carried by a quantum of radiation. In chapter 7 (section II.7.4), we will have more to say about the "recoiling electrons", predicted by Compton's theory.

[6] A result reached empirically by Edwin Augustine Owen. See [Owen (1918)], p. 522.

Chapter II.5

Solvay IV (1924): Electric conductivity in metals

Solvay IV

Courtesy of the Solvay Institutes

Ernest Solvay died in May 1922, but his *Science Project* was alive and about to expand to other regions of the world. In 1923, only five European countries were represented in ISIP's Scientific Committee.[1] Fortunately, some of its members (in particular Lorentz) had established close contacts with physicists working outside Europe.

The fourth Physics Council was to be held in the spring of 1924. Plans for this conference were made by the ISC during its meetings of April 18 and 19, 1923. Besides the difficulty of discussing progress in quantum theory in the absence of German researchers, the timing for such discussions was hardly favourable. Major advances that were to mark the last months of the "Old Quantum Theory" had not yet taken place. Hendrik A. Kramers's *dispersion relation*, a crucial source of inspiration for Heisenberg, and Wolfgang Pauli's *exclusion principle* would only appear in 1924. As for the electron *spin*, it had still to be introduced by George Uhlenbeck and Samuel Goudsmit in a letter of November 20, 1925. See [Uhlenbeck and Goudsmidt (1925)]). As noted by Pais ([Pais (1991)], p. 241):

> The letter contained the correct interpretation of Sommerfeld's hidden rotation, which earlier that year had been traced by Pauli to an "intrinsic two-valuedness" of the electron ... The work of Uhlenbeck and Goudsmit was entirely in the spirit of the old quantum theory, even though the new quantum mechanics had already arrived a few months earlier.

II.5.1 Subjects to be discussed at Solvay IV

Rutherford and van Aubel, who were not present at the meeting, sent their proposals by letter. The Committee members agreed to select a subject that could be dealt with in the absence of German participants.[2] Four topics were proposed: *quanta, electric conductibility, phenomena at very low temperatures, the theory of relativity*. After lengthy discussions, the chosen topic was "The mechanism of metallic conductivity and a series of phenomena likely to shed light on that mechanism".

[1]The ISC comprised the following members: Lorentz, Marie Curie, Brillouin, Rutherford, W. H. Bragg, Kamerlingh Onnes, Langevin, Knudsen, and van Aubel (Langevin had been chosen in 1921 to fill the seat of Righi, who had died in June 1920).

[2]Lorentz made clear that the rapporteurs would be authorised to cite German papers.

The subject was of particular interest to Lorentz because of its relationship with the Theory of Electrons, and the persistence of a mystery: "Why did electrons, the recognised carriers of electricity and heat in metals, not contribute any part in the specific heat of a solid?"

As noted by André Langevin ([Langevin (1966)], p. 13), the theme for Solvay IV was well chosen because it allowed the invitees to discuss the *application of quantum theory* to specific problems such as the change in metallic conductivity at very low temperatures, and the phenomenon of superconductivity, topics that could be addressed by non-German physicists. The researchers Lorentz had in mind comprised Kamerlingh Onnes, Peter Debye, Willem H. Keesom, Lindemann and George de Hevesy. André Langevin emphasised the connection with the theory of quanta and his father's suggestions. Paul Langevin insisted to invite Léon Brillouin – son of Marcel Brillouin – who, according to André Langevin ([Langevin (1966)], p. 13), should have presented for the first time a report where he applied the quantum theory to the electronic theory in metals. André Langevin's preliminary list[3] of reports and rapporteurs looks as follows ([Langevin (1966)], p. 13):

(i) Metal conductivity (except superconductivity). Rapporteurs: Lorentz and Percy Williams Bridgman, from *Jefferson Physical Laboratory* (MA – USA).
(ii) Superconductivity. Rapporteurs: Kamerlingh Onnes and Paul Langevin.
(iii) Conductibility in crystals. Rapporteurs: Joffé, professor at the *Polytechnic Institute of Leningrad* (Russia).
(iv) Magnetic field effects. Rapporteurs: Edwin Hall, Professor at the *University of Cambridge* (USA), and Léon Brillouin, Doctor of Science (Paris).

However, as we will see in section II.5.2, Léon Brillouin did not present a report; the last contribution was presented by Hall.

In April 1923, Lorentz may have had a personal reason for selecting the problem of metallic conductivity. The idea was apparently connected to his meetings with American and Canadian physicists during his transatlantic tour from January to April 1922, a journey which enabled him to meet Hall, the discoverer of the phenomenon that bears his name, Bridgman, a specialist in high pressures; John MacLennan, a low-temperature expert

[3]For his article, it is likely that André Langevin use his father's personal notes.

from Toronto, and Compton, the expert on secondary radiations produced by X-rays, who prepared the report we commented on in section II.4.2.

II.5.2 Invited members and reports

The Council was expected to take place from April 24 to 29, 1924. In view of the above considerations, it is not surprising that the list of guests, drawn up in April 1923, comprised the names of Bridgman, Debye, Keesom*, Lindemann, de Hevesy*, MacLennan*, next to those of Bohr, Ehrenfest, Erwin Schrödinger, Léon Brillouin, Orso Mario Corbino, Joffé,[4] Władisław Broniewski, Richardson*, W. Lawrence Bragg*, Albert W. Hull,[5] de Haas*.[6]

An invitation was also extended to Einstein, in agreement with the Institute's administrators. However, Einstein declined the invitation, arguing that he no longer wanted to be the only German member.[7] As a contrast, the Council welcomed Schrödinger, an Austrian physicist working at the University of Zurich, and de Hevesy, a Hungarian researcher attached to Bohr's Institute in Copenhagen. Their presence at Solvay IV was justified by their expertise on the role of electrons in metals and crystals:

- Schrödinger was known for his interest in the free electron gas in metals, a test case for *Gasentartung* (a precursor for what would become quantum statistics). He had recently published a paper on the subject.[8]
- De Hevesy, who had been close to Moseley at Rutherford's laboratory, was known for his work on the conductivity of crystals, and his research on the change in conductivity of molten salts at solidification.[9]

The participation of these physicists in the Council offered the advantage of reducing the impression of a permanent boycott of researchers belonging to the former Central Powers.

---

[4] Joffé had spent two years in Munich as an assistant to Wilhelm Röntgen. In 1911, Joffé had determined the charge of the electron, independently of Millikan.

[5] In section II.7.4, we will see that Hull played a role with respect to A. H. Compton's publication in 1923 of his first report on the scattering of X-rays by free electrons.

[6] The symbol (*) indicates that these physicists were "kept in reserve".

[7] Einstein even begged the Institute not to invite him again. It is likely that his reaction was due, in part, to the exorbitant sums imposed on Germany by the Allies, and the occupation of the Ruhr by French and Belgian troops. See [Born (2005)], letters 29 and 45.

[8] See [Schrödinger (1924)]. The authors thank Michael Eckert for providing them with this information. Further details can be found in [Eckert (1989)].

[9] See [Cockcroft (1967)]. De Hevesy was elected Foreign Member of the Royal Society in 1939.

It is worth emphasising that in April 1924 Schrödinger had not produced his celebrated *wave mechanics*, a theoretical formalism based on ideas developed by Louis de Broglie. His invitation to attend the Council, along with de Hevesy, illustrates the Committee's ability to detect the merits of a physicist before he produced the result that would make him famous (Schrödinger would get the 1933 Nobel Prize in physics, along with P. A. M. Dirac, for his work on quantum mechanics; de Hevesy would be awarded the 1943 Nobel Prize in Chemistry for his pioneering work on radioactive tracers).[10]

Another noteworthy point was the invitation of the Ukrainian physicist Abram Fedorovich Joffé, an expert in solid state physics, considered today as the father of Soviet Physics. According to André Langevin ([Langevin (1966)], p. 13), Joffé was invited at Paul Langevin's suggestion.[11] Regarding Russia, we should mention that Ehrenfest, who spent five years in St. Petersburg before becoming Lorentz's successor in Leiden, was prevented from taking part in the Council because of a visit to the USA. Lorentz reported his absence in a note that was published in the newspaper *L'indépendance Belge*: "A trip to America prevented Professor P. Ehrenfest, from Leiden, from going to Brussels, where his absence was greatly regretted by his colleagues".

When the Scientific Committee met in April 1924, the Council's programme had been significantly expanded:

1. *Application of electron theory to metals and their properties* (Lorentz).
2. *Conductivity phenomena in metals and their theoretical explanation* (Bridgman).
3. *A theory of conductivity in metals* (Richardson).
4. *The internal structure of alloys* (Walter Rosenhain).
5. *Electrical resistance and metal expansion* (Broniewski).
6. *Electrical conductivity of crystals* (Joffé).
7. *New experiments with superconductors* (Keesom[12]).

[10]De Hevesy's discovery in 1922 of "hafnium" by means of X-ray spectroscopy, together with the Dutch physicist Dirk Coster, may also have played a role in his invitation. Having noticed that a chemical element with 72 protons was missing in Mendeleev's periodic table, de Hevesy had been looking for the element on the basis of Bohr's atomic model.

[11]André Langevin argued that his father was impressed by Joffé's work on semi-conductors properties, which opened the creation of the *transistors*.

[12]Willem Keesom replaced Kamerlingh Onnes, who had fallen ill.

8. *Metal conductivity and the transverse effects of the magnetic field* (Hall).
9. *Note on the propagation of radiation impulses* (Joffé and Dobronravoff).

II.5.3 Solvay IV and quanta

Despite the absence of German members, the Council discussed some specific quantum-theoretical questions. This was notably the case during the discussion of the reports by Bridgman and Richardson, in which references were made to Bohr's theory with regard to the distribution of atomic electrons. Bohr's "second" atomic model was addressed by Richardson in his theory of electrical conductors, based on "current theories of photo-electric action and Bohr's spectra", and also by Keesom, in his presentation of Kamerlingh Onnes's report on "Superconductors and the Rutherford–Bohr Atomic Model". Keesom showed diagrams with orbits in multi-electron atoms, taken from a recently published paper by Kramers that established a connection between the dispersion of radiation by atomic systems and Einstein's probabilistic formulation of radiative processes.

In his discussion of the phenomenon of superconductivity, Richardson referred to the Stern–Gerlach experiment of 1922 and its result suggestive of "spatial quantisation" – a discovery high-lighted by Einstein[13] as "the most interesting thing at the moment."

Another notable result, although less known, was the one presented by Joffé on the quantum nature of light. The idea of asking Joffé to present a "Note" on his recent experiments with Dobronravoff, had been suggested, presumably by Lorentz, a few days before the meeting. While visiting Leiden before the Council, Joffé had informed his Dutch colleagues of the results he and Dobronravoff had obtained regarding the energy distribution in X-rays. Impressed by their findings, Lorentz had expressed enthusiasm in a letter to Ehrenfest ([Kox (2018)], p. 261): "This is certainly the nicest combination of new discoveries that I know of."

The point to be emphasised is that Joffé referred in his presentation to Maurice de Broglie's results, showing that the energy received by secondary electrons, expelled by X-rays, is equal to the energy needed for a cathode

---

[13]See Einstein's answer to a letter from Max Born of April 30, 1922 ([Born (2005)], letter 42). In a comment Born complained: "It is strange that Einstein referred me to Stern's and Gerlach's experiment as the most interesting. He had forgotten that it had been carried out in my Institute under my very eyes, as a result of discussions with me". In fact, the Stern–Gerlach experiment, the prototype of quantum measurement, was to be considered a "German" exploit.

electron to create these rays. Aware of the fact that this observation could also be explained by an "accumulation" of energy by the secondary electron, at the expense of a large number of separate Röntgen pulses, Joffé thought of an experiment that could exclude this possibility. Using Millikan's method, he came to the following conclusion:

> We must admit that the energy of a single pulse is concentrated at a precise point and is not distributed throughout the entire angle: this is Einstein's light quantum hypothesis. If we want to recognise the law of conservation of energy in an elementary process such as a transformation of radiant energy into the energy of an electron, our experiments, put in relation to those of Mr. de Broglie and Mr. [Richard] Widdington, lead us to conclude that the energy of a single pulse is always emitted in one place only. Our results are in agreement with a similar case observed by Mr. Wilson and published in the August–September 1923 issue of the Proceedings of the Royal Society.[14]

As we will see, soon after Solvay IV (May 1924), Bohr, Kramers and John Slater published a paper that contained the so-called BKS proposal, based on the idea that conservation of energy would not apply to elementary processes, but only hold as a statistical principle.[15] The contrast between Joffé's conclusion and the BKS proposal, to which also Born adhered,[16] is representative of a difference between the early stages of the two lines of research that led to the birth of quantum mechanics: the *Schrödinger line* and the *Heisenberg line*.

II.5.4 Schrödinger and Louis de Broglie's doctoral thesis

In 1962, Joffé published a book in which he recounts his memories of the fourth Physics Council. In this book *Encounters with physicists* published in Moscow and translated in German in 1967, Joffé reports certain facts

[14]From 1908 to 1914, Richard Widdington was a research student of J. J. Thomson with whom he mainly worked on cathode rays (see Cavendish Laboratory: *Portrait of Richard Whiddington*). Heilbron identifies Widdington's 1911 observation about the threshold electron energy needed to excite a given X-ray as the key empirical insight that guided Bohr to adopt nuclear charge as defining an atom's identity ([Friedrich (2019)]). During the First World War, Widdington became Cavendish Professor of Physics at Leeds University (1919–1951).

[15]See Appendix B for some further details.

[16]See [Heilbron and Rovelli (2023)], p. 4.

which took place during the meeting.[17] He also provides indications that it was at Solvay IV that Schrödinger learned of the existence of Louis de Broglie's doctoral thesis. This point was briefly mentioned by Jammer ([Jammer (1989)], p. 251):

> Now, as we know from reliable sources, Langevin, in spite of any as yet recognised empirical verification, spoke about de Broglie's work at the fourth Solvay Congress in April 1924. He also informed Einstein about it. Einstein, working at that time on the implications of [Nathan] Bose's paper [Bose (1924)], became naturally very interested in de Broglie's thesis and asked for a copy, which he received and read in December 1924. As a result, in his second paper on the quantum theory of the ideal gas, Einstein referred, in the context of his discussion on the interference term in the energy-fluctuation formula, to de Broglie's work and declared: "I shall discuss this interpretation in greater detail because I believe that it involves more than a mere analogy."

One of Jammer's reliable (yet unknown) sources was presumably an excerpt from the text written by Joffé on the occasion of Langevin's death in December 1946 ([Joffé (1947)], p. 15):

> I remember a meeting with Paul Langevin in 1924 at the Solvay Congress, where the contradictions of classical and atomic physics manifested themselves in a particular clear way. Lorentz, the creator of the electron theory, could only express his regret at not having died five years earlier, when the horizon of his theory still seemed serene. Langevin spoke of Louis de Broglie's thesis as an interesting attempt at synthesis, and a possible way out of the impasse.

In his book *Encounters with physicists*, Joffé also refers to Langevin's reaction to de Broglie's thesis:

> In the spring of 1924, in Brussels, Langevin told me about Louis de Broglie's ideas about electron waves. He submitted the work for a doctorate award to de Broglie. Langevin was charmed by the insight and originality of de Broglie's ideas, even if he did not believe they were real.

[17]See Appendix A.

In any case, there is a point that should be emphasised: the most lively discussions at Solvay meetings often took place outside the conference room. Joffé explains that during the breaks he spent a lot of time with Debye and Schrödinger. The latter had several opportunities to discuss recent events with Langevin. Hence, we have reasons to believe that, during one of their exchanges, Langevin informed Schrödinger about de Broglie's doctoral thesis. On the other hand, there are indications that Schrödinger did not follow up on this information, and only took notice of de Broglie's publications in 1925.

In 1964, Debye revealed, in an interview,[18] the circumstances that led Schrödinger to start his celebrated work ([Jammer (1989)], pp. 257–258):

> Then de Broglie published a paper. At that time Schrödinger was my successor at the University of Zurich, and I was at the Technical University, which is a Federal Institute, and we had a colloquium together. We were talking about de Broglie's theory and agreed that we did not understand it, and that we should really think about his formulations and what they mean. So, I called Schrödinger to give us a colloquium, and the preparation of that really got him started. There were only a few months between his talk and his publications.

Jammer continues as follows:

> Debye's choice, it seems, was not accidental. For Schrödinger – and this brings us to a second and more profound incitement – was very eager to study de Broglie's ideas and their possible implications. Interested in statistical mechanics, he had read Einstein's paper on the quantum theory of an ideal gas, in which ... Einstein expressed his belief that de Broglie's conceptions "involve more than merely an analogy". In fact, it was this remark which induced Schrödinger to study what he referred to at that time as the "de Broglie-Einstein" undulatory theory, according to which a moving corpuscle is "nothing but the foam on a wave of radiation in the basic substratum of the universe".

[18] The original interview dated 1964 can be found in *Science* **145** (pp. 554–559). We chose to reproduce the part of the interview cited by Jammer as well as Jammer's comments on the subject.

Jammer also reports a "French story" about Schrödinger's reaction to reading de Broglie's thesis:

> There are, however, other accounts for which the author was unable to find additional independent evidence: Thus, Victor Henri, a former pupil of the physiologist Albert Dastre of the Sorbonne, and since 1924 professor of physical chemistry at the University of Zurich, recalled – according to Edmond H. Bauer (Archive for the History of Quantum Physics, Interview with E. Bauer on January 8, 1963) – that, while visiting Paris, he received from Langevin a copy of the "very remarkable theory of de Broglie"; back in Zurich, and having not well understood what it was all about, he gave it to Schrödinger, who after two weeks returned it to him with the words: "That's rubbish". When visiting Langevin again, Henri reported what Schrödinger had said. Whereupon Langevin replied: "I think Schrödinger is wrong; he must look at it again". Henri, having returned to Zurich, told Schrödinger: "You ought to read de Broglie's thesis again; Langevin thinks this is a very good work". Schrödinger did so and began his work.

A brief, but illuminating summary of Louis de Broglie's "revolutionary" proposal has been given by Pais ([Pais (1991)], p. 240):

> Where Bohr wanted to exclude light particles and stick to waves, de Broglie, inspired by Einstein's wave-particle duality for light, proposed that matter too, specifically electrons, should under suitable circumstances, not behave like the familiar particles but like never yet suggested waves.

According to John C. Slater, it took some time for de Broglie's ideas to spread to the main research centres. As noted by Pais, Slater made this clear in his scientific biography ([Slater (1975)], p. 12): "None of us in Copenhagen in the spring of 1924 had known of the work of de Broglie in Paris ... His main publications did not reach attention until early 1925 ... Even in 1925 they were not common knowledge."

As for Schrödinger, he left no doubt about the roots of his ideas ([Schrödinger (1926c)], p. 734):

> My theory was stimulated by de Broglie and brief but infinitely far-seeing remarks by Einstein. I am not aware of a generic

connection with Heisenberg. I, of course, knew of his theory but was scattered away, if not repulsed, by its transcendental algebraic methods which seemed very difficult to me.

Speaking of his *wave mechanics* in a letter to Einstein of April 1926, Schrödinger declared ([Jammer (1989)], p. 258): "The whole business would not have been produced now, and perhaps never (I mean not by me) if one of your papers had not impressed me with the importance of de Broglie's ideas."[19]

Louis Victor de Broglie (1892–1987)
Courtesy of the Fondation Louis de Broglie

[19]Schrödinger referred to [Einstein (1925)].

Chapter II.6

The "Heisenberg line" of research

Paul Adrien Maurice Dirac (on the left) walking with *Werner Heisenberg* (on the right) through *Solbosch Campus* at the Université Libre de Bruxelles. The picture also shows the place where the Solvay Archives are kept nowadays.

Courtesy of the Niels Bohr Archive, Copenhagen

II.6.1 The Göttingen school

In his comments on letters exchanged with Einstein ([Born (2005)], p. 58), Born reported on the beginnings of his Göttingen work on atomic structure and the theory of atomic spectra. Referring to a letter to Einstein, dated October 21, 1921 ([Born (2005)], letter 33), he made the following remark about the collaboration he had started with his young assistant Pauli:

> The aim was to see whether the Bohr–Sommerfeld rules for the application of the quantum hypothesis to mechanical systems would lead to correct results. We used appropriate approximation methods based on Poincaré's astronomical perturbation calculation.
>
> The results were negative ... Soon after, however, I saw it differently. Was it possible that the success of Bohr's theory in the case of hydrogen and other similar systems was a sort of accident? Maybe there was a better theory? This became our research programme, especially after Heisenberg succeeded Pauli: we began to look systematically for cases where Bohr's theory failed.

Applying a perturbation scheme, Born and Heisenberg found in 1923, that Bohr's theory failed in the case of the Helium atom.[1] This result convinced them of the need to turn away from classical mechanics and set up a new theory: "quantum mechanics". A first, but decisive step in this direction was the manuscript Heisenberg handed over to Born in July 1925, after his "productive sojourn" in Helgoland ([Heilbron and Rovelli (2023)], p. 5).

II.6.2 Dirac recounting the birth of quantum mechanics

Fifty years after Heisenberg's publication of his first paper on quantum mechanics [Heisenberg (1925)], Paul Adrien Maurice Dirac, another builder of the new mechanics, delivered a speech at the New South Wales Branch of the Australian Institute of Physics [Dirac (1978)]. In this "illuminating" report he recalled the way in which Heisenberg was led to his revolutionary ideas, and how he himself succeeded in transforming Heisenberg's theory into an elegant formalism based on the use of operators[2]:

[1] See [Born (2005)], letter 45.
[2] The paper Dirac was referring to is [Dirac (1925)].

Quantum Mechanics is a development of the classical mechanics of Newton ... According to Bohr we have to assume that the atom can exist only in certain states called stationary states. These states are fixed by some conditions of a type which is entirely unfamiliar to classical mechanics: a set of new conditions called quantum conditions, which involve Planck's constant h.

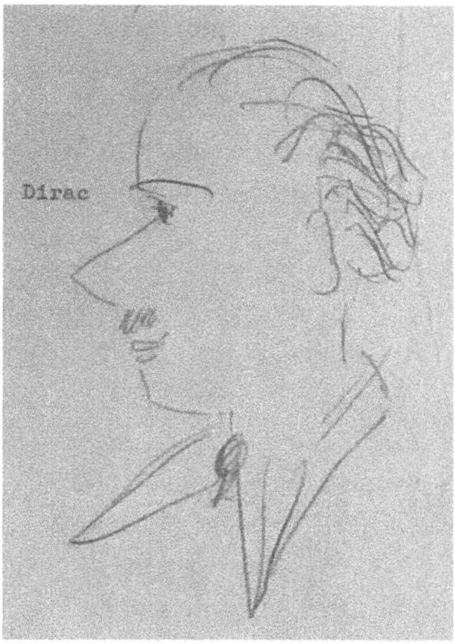

Paul Adrien Maurice Dirac as pictured by one of the participants of a later Solvay Conference
Courtesy of the Solvay Institutes

An atom, according to Bohr, can jump from one stationary state to another, and, when it jumps, it emits (or absorbs) radiation to preserve conservation of energy. This radiation, which is emitted or absorbed, is a single quantum with a definite frequency connected with the energy ...

I believe that the introduction of these ideas by Bohr was the greatest step of all in the development of quantum mechanics. It is really the most unexpected, the most surprising thing that such a radical departure from the laws of Newton should be successful ...

Well, people worked on this theory of Bohr, and they found that the successes were very limited. Essentially, one had successes when one was dealing with an atomic system for which there was just one electron playing an important part. If one had two electrons, or more than two, then one did not know how to apply the quantum conditions to fix the stationary states.

People did various calculations, making artificial assumptions, but the work was not successful ... The great advance was made by Heisenberg in 1925. He made a very bold step. He had the idea that physical theory should concentrate on quantities which are closely related to observed quantities. Now, the things you observe are only very remotely connected with the Bohr orbits. So, Heisenberg said that the Bohr orbits are not very important. The things that are observed, or which are connected closely with the observed quantities are all associated with two Bohr orbits and not with just one Bohr orbit.

Now, what is the effect of that? Suppose we consider all the quantities of a certain kind associated with two orbits, and we want to write them down. The natural way of writing down a set of quantities, each associated with two elements, is in the form of an array of quantities which one sets up in terms of rows and columns. One has the rows connected with one of the states, the columns connected with the other. Mathematicians call a set of quantities like this a matrix. Now Heisenberg assumed that we should deal with such a set of quantities and consider the whole set together as corresponding to one of the dynamical variables of the Newtonian theory. These dynamical variables are, of course, the coordinates of the particles, or the velocities or momenta. Each of these quantities is to be replaced by a matrix, according to Heisenberg.

The underlying idea that led Heisenberg to think of this was that one should construct the theory in terms of observable quantities and that the observable quantities are these matrix elements, each associated with two orbits ... Actually, the most important point of the new mechanics of Heisenberg is that dynamical variables are subject to an algebra where we do not have commutative multiplication. Well, from the initial idea of Heisenberg, one could make a fairly rapid development, and I was able to join in ... We had the problem, you see, of changing the equations of Newton so as to fit in with an algebra where "a times b" is not equal to "b times a". Now, at first sight, it might seem that that would be rather difficult. But the problem was made very much simpler by some work of Hamilton which was done a hundred years previously. Hamilton had been studying the equations of Newton, and he found another way of writing them ...

The importance of the Hamiltonian form of the equations is that one can very easily generalise them to bring in the non-commutation. One can write the Hamiltonian equations in terms of certain bracket expressions called the Poisson[3] brackets ... By replacing the Poisson brackets of the Hamiltonian form by the commutator a times b minus b times a, we can immediately pass from all the equations of classical mechanics (after we have put them in Hamiltonian form) to some new equations, where we have non-commutative multiplication, and which we can apply to the Heisenberg form in quantum mechanics ...

We had then equations involving these non-commuting quantities, but to begin with, we had no general interpretation for the equations. In any physical theory one usually knows what one's equations mean before one sets them up. But here, we had a different situation: we had equations before we knew how to apply them...

II.6.3 Early reactions to Heisenberg's theory

From the end of 1925, Einstein commented on the new mechanics. In November 1925, Bose noted ([Pais (1991)], p. 317): "Einstein was very excited about the new quantum mechanics. He wanted me to try and see what the statistics of light-quanta and the transition probabilities of radiation would look like in the new theory". However, as Pais recalls ([Pais (1991)], p. 322, Note 80):

During the following months Einstein seems to have vacillated in his reactions to the matrix theory. In December he wrote to his friend Michele Besso: "The most interesting theoretical achievement at the moment is the Heisenberg–Born–Jordan theory of quantum states.

A real sorcerer's multiplication table, in which infinite determinants (matrices) replace the Cartesian coordinates. It is extremely ingenious, and thanks to its great complexity sufficiently protected against disproof.

[3]The brackets were introduced by Simenon–Dénis Poisson.

Reactions to Heisenberg's theory came also from Holland. In a letter of January 5, 1926, Ehrenfest told Lorentz ([Kox (2018)], doc. 272): "I started studying matrix mechanics with Karel M. Nuyens, one of [Théophile] De Donder's students." On January 11, Lorentz replied:

> Yes, matrix mechanics is a difficult subject; an "equation of motion", that has lost all its usual meaning, but is still valid! I will also see what I can make of it, when I am back in Leiden.

Meanwhile Schrödinger's first papers on wave mechanics had appeared. In a letter of April 26, 1926, Einstein told Schrödinger: "I am as convinced that you have made a decisive advance with your formulation of the quantum condition as I am that the Heisenberg–Born route is a wrong track."[4]

II.6.4 Dirac about Schrödinger

Let us return for a moment to Dirac, who in his *Lectures* at the University of New South Wales, emphasised the role played by Schrödinger:

> In any physical theory one usually knows what one's equations mean before one sets them up. But here, we had a different situation: we had equations before we knew how to apply them... Well, we had to face the problem of getting a general interpretation for these new equations, and this problem was very much assisted by some work of Schrödinger.
>
> Schrödinger was working independently of Heisenberg, and set up a theory of his own, an alternative form of quantum mechanics, which at first sight appeared to be completely different from the Heisenberg theory. Schrödinger came a bit later than Heisenberg. After a few months it was established that the Schrödinger theory and the Heisenberg theory were really equivalent, although they looked very different in their starting points.
>
> Schrödinger's theory was based on some previous work of [Louis] de Broglie, who showed how one could introduce waves connected with particles. De Broglie had wave functions which are usually written "psi". For a single particle this "psi" would be a function of the coordinates of the particle (including time). De Broglie

[4]As noted by Pais, Einstein's letter to Schrödinger is also reproduced in [Klein (1967)].

set up an equation to govern the waves described by "psi", and this wave equation is such that if one takes plane waves moving in a definite direction, with a definite frequency, they correspond to a particle with a definite momentum and definite energy. The correspondence is a relativistic correspondence and is a very neat one mathematically.

De Broglie was led to this idea of connecting waves with particles by mathematical beauty. De Broglie's theory applied only to particles not acted on by any forces. But Schrödinger was able to generalise the theory so as to apply it to an electron moving through an electromagnetic field, acted on by electric and magnetic forces.

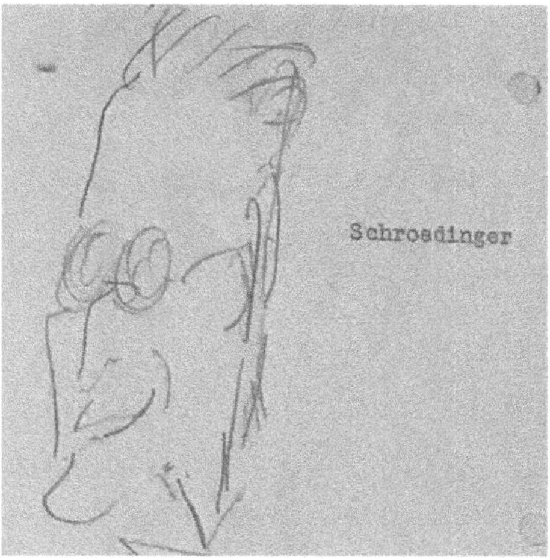

Erwin Schrödinger as pictured by one of the participants of a later Solvay Conference
Courtesy of the Solvay Institutes

In Schrödinger's theory, one had operators acting on "psi" ... and each of these operators corresponds to one of the momentum operators. If one is dealing with such operators, then one is dealing with non-commuting quantities that are similar to the non-commuting quantities of the Heisenberg theory. And on that basis, one is able to establish the connection between the Schrödinger theory, involving these wave functions, with the operators operating on them, and the Heisenberg theory. Now, the "psi" itself did not occur in the original Heisenberg form of the theory, but it got introduced into quantum mechanics through the work of Schrödinger. And then one found that the wave function "psi" corresponds to one of the states, perhaps one of the stationary states of the Bohr theory.

The operators, which change one of the wave functions to another, are thus connected with two states. And so, the equivalence of Schrödinger's wave mechanics and Heisenberg's matrix mechanics was established.

Chapter II.7

Towards Solvay V

H. A. Lorentz
Courtesy of the Solvay Institutes

II.7.1 1925: Lorentz's jubilee

The year 1925 marked the fiftieth anniversary of Lorentz's doctoral dissertation. Frits Berends, Emeritus Professor of the University of Leiden, published a booklet entitled[1] *Lorentz's golden doctorate: an unprecedented tribute*, for the book series *Stories around a photo.*[2] Berends reconstructed the event and its historical framework.

The celebration was organised by Kamerlingh Onnes, Zeeman, and Adriaan Fokker ([Berends (2014)], p. 37). The picture below shows some participants,which included Marie Curie, Niels Bohr, Arthur Eddington, Willem Keesom, Jean Perrin, Paul Langevin, Paul Ehrenfest, and Albert Einstein, surrounding Lorentz and his wife.

Lorentz Jubilee in Harlem
Courtesy of the Solvay Institutes

During the event that took place on December, 11, 1925, Marie Curie expressed her admiration for "the extreme care with which Lorentz

[1]The original title in Dutch reads: *Lorentz' gouden doctoraat: een huldiging zonder weerga.*
[2]In Dutch: *Verhalen rond een foto.*

prepared the Solvay Councils" ([Berends (2014)], p. 49). According to Berends, she did not know what to value most in his presidency: his courtesy, kindness, linguistic skills, clarity of mind, or the flame of enthusiasm that made him the youngest of physicists, and simply concluded that Lorentz was *the true animator of the Solvay Councils.*

Marie Curie also spoke on behalf of the International Commission on Intellectual Cooperation,[3] born within the *League of Nations*, expressing her thanks for Lorentz's valuable participation. In 1925, Lorentz had already started working for readmitting German scientists to the international scene. As we have seen in section II.2.1, he supported the idea of keeping the door open to German scientists after a period of gradual reconciliation.

Speaking on behalf of the scientists of many countries, Einstein underlined Lorentz's efforts ([Berends (2014)], p. 49):

> You suffered like few others from Europe's misfortune and you did everything in your power to bring about united cooperation again. That is why we love and admire you as a person and not just as a researcher.

In 1925, Lorentz did not succeed in bringing about the expected result in the context of the IRC.[4,5] As we will see, his efforts were successful within the narrower circle of the International Solvay Institute.

II.7.2 Launch of the fifth Physics Council

The members of the ISC met in Brussels on April 1 and 2, 1926, to fix the programme of Solvay V. Their first concern was to decide who would succeed to Kamerlingh Onnes, who had died a few weeks earlier. Langevin ([Langevin (1966)], p. 11) suggested giving his seat to Einstein. The proposal was unanimously approved. The Committee members also agreed that is was time to invite "*moderate*" German physicists. The Administrative Commission accepted the idea.

---

[3]See [Berends (2014)], p. 35.

[4]After the end of WWI, the Allied academies had cancelled their membership in the International Association of Academies before creating a new organization, the *International Research Council* (IRC).

[5]See footnote 4.

The agreement was facilitated by Germany's decision to join the League of Nations.[6] However, the presence in Brussels of German citizens remained a sensitive issue, which required royal approval. On April 2, Lorentz was received in audience by King Albert,[7] who declared:

> A better understanding between peoples is absolutely necessary and science should help bringing it about. Considering what the Germans have done in physics it would be difficult to do without them.

That settled the problem.

The Committee then turned to the actual preparation of Solvay V. In anticipation of the meeting, Lorentz had sought advice from Ehrenfest, his successor in Leiden, who used to be well informed about the latest advances in physics. On Monday, March 29, 1926, Lorentz sent the following message to Ehrenfest from Paris[8]:

> Next Thursday we will plan the Solvay meeting for next year. Among the subjects to be dealt with, there is the"conflict and possible reconciliation between classical theories and quantum theory" (certainly the most up-to-date subject of all). I have been thinking of physicists we could invite (including you, but you should first give me an answer), but fear I have forgotten some.
>
> This is a list of names that came to my mind: Einstein, Bohr, Kramers, Born, Heisenberg, Jordan (but he is a mathematician), Pauli, [Rudolf] Ladenburg (?), Slater, the young Bragg (because of the agreement his work produced with classical theory), J. J. Thomson, and one or two other Brits (Darwin,? Fowler ?), Léon Brillouin (I do not know if he has worked on the subject, he was here several times), Louis de Broglie (light quanta), one or two researchers who have been concerned with the refraction of X-rays (Bergen Davis?, Compton, Debye, Dirac ?).
>
> Would you be so kind as to add some more names to this list, and give me your opinion on those that I mentioned with a question

[6]See the telegram sent by Jules Bordet, King Albert's representative in the ISC, to Lorentz on April 2, 1926 (S.a.b.ULB).

[7]Letter sent by Lorentz to [Charles] Lefébure, dated April 3, 1926. S.a.b. ULB, doc. 828.

[8]Letter from Lorentz to Ehrenfest, March 29, 1926, Haarlem, Noord-Hollands Archief, Archief H. A. Lorentz, (Archive 364).

mark? I would be very pleased if I could receive your answer at my address in Brussels on Wednesday evening.

Ehrenfest was not part of the International Scientific Committee, but he certainly was a well-informed physicist whom Lorentz used to consult on current affairs. At the end of his letter, Lorentz indicated his address in Brussels for the ISC meeting – Fondation Universitaire, rue d'Egmont, 11 – because he was hoping to receive Ehrenfest's answer in two or three days, i.e., in time for the ISC gathering.

It is interesting to note that Lorentz did not mention Schrödinger. Ehrenfest, on the other hand, scribbled Schrödinger's name, together with other names on the last page of Lorentz's letter. He suggested the following order:

1. Ralph Fowler
2. Dirac
3. Franck
4. Enrico Fermi
5. Darwin
6. Schrödinger
7. Oseen
8. Walther Bothe
9. Adolf Smekal

On March 30, Ehrenfest sent a letter to Lorentz,[9] written late at night, which started as follows:

> Since this letter may come too late, I will telegraph to you tomorrow. Nevertheless, here is a hasty reply. I content myself with making some individual remarks on the order of priorities (which seems to be adapted to the purpose of the discussion) and on some other considerations.

In the letter, Ehrenfest mentioned the above names, together with those of Langevin and Léon Brillouin, some with a brief commentary:

> Schrödinger, could perhaps be invited as the first who gave a quantum interpretation of the Doppler effect also close to the Compton effect.

[9]Haarlem, Noord-Hollands Archief, Archief H. A. Lorentz (Archive 364). The authors are grateful to Frits Berends for providing them with a copy of the letter.

[Walter] Bothe, because of the Bothe–Geiger experiment which destroyed the Bohr–Slater theory[10] – he might even be considered before Schrödinger.

However, Ehrenfest strongly criticised the Council's "theme" that had been proposed by Lorentz:

> The slogan "Possible reconciliation between classical theories and the theory of quanta" appears to me, perhaps wrongly, as extremely unpleasant. It would cause many minds to evade the fruitful and suggestive harshness of the conflict by thinking as slyly and unclearly as possible, just as happened after 1900 with mechanical aether theories in the case of Maxwell's equations. I do not dare to push my taste too far in this direction, but I know that Bohr is even more opposed to the slogan than I am, precisely because he is so keen to find the foundations of the new theory.

The Lorentz–Ehrenfest exchange sheds light on several points. By March 29, 1926, Lorentz still believed in a possible reconciliation between quantum theory and the existing theories. This hope, especially regarding the classical theory of light, to which Lorentz contributed so brilliantly, may have been strengthened by Planck's words in a letter[11] of December 15, 1925:

> After many disappointments, I now hope to experience the integration of quantum theory into a unified world view. The latest investigations by Born and Jordan on matrix physics seem to be suitable for overcoming the antagonism between the differential equations of classical theory and the quantum theory, since the spherical wave is retained to a certain extent, and thus the interference phenomena. This unsolvable riddle of the quantum concept of light becomes understandable. But the matter is not yet certain.

Before the April meeting of the ISC, Ehrenfest and Lorentz were both unaware of Schrödinger's groundbreaking work.[12] Ehrenfest suggested

[10]See section II.7.4 for more information on what Ehrenfest called the "Bohr–Slater" theory.

[11]Letter from Planck to Lorentz, Haarlem, Noord-Hollands Archief, Archief H. A. Lorentz (Archive 364).

[12]Schrödinger's first paper on wave mechanics was received by *Annalen der Physik* on January 27, 1926.

to invite Schrödinger because of his paper of 1922, published in the *Physikalische Zeitschrift.*[13] As for Lorentz, he would only be informed of Schrödinger's recent work after the ISC meeting.[14] Yet, remembering Schrödinger's presence at Solvay IV, he welcomed Ehrenfest's suggestion and decided to ask Schrödinger to report on *Adaptation of the foundations of dynamics to quantum theory.*[15]

Lorentz also accepted Ehrenfest's criticism of a "possible reconciliation of quantum theory with classical theories", as shown by the Committee's decision to devote Solvay V to "The theory of quanta and the classical radiation theories". A fortnight earlier, Lorentz had complained in a letter to Einstein ([Kox (2009)], doc 404): "I still can't get to grips with the matrix mechanics". On April 6, after the ISC meeting, he wrote again to Einstein, informing him of the theme of the Council, along with the other decisions that had been taken by ISIP ([Buchwald *et al.* (2021)], doc.246). The following lines are part of Lorentz's letter:

> Now I have to write to you about another matter. From Paris, I went to Brussels, where the scientific Committee of the *Institut International de Physique Solvay* met with the purpose of preparing the next congress of physics, which is supposed to be held in October 1927. The above-mentioned Committee, composed of Mrs. Curie, Langevin, Richardson, [Charles-Eugène] Guye (Geneva), Knudsen, van Aubel (Ghent) and me, is now in session. Only Bragg and van Aubel were absent. We proposed to the Administrative Commission that you be appointed as a Committee member and a successor to Kamerlingh Onnes. The Administrative Committee immediately accepted the idea, and it would now please all of us very much if you would like to occupy this place in the Committee. Because I know that much is demanded of you from every quarter, I add that now it does not involve a great expenditure of time at all. A congress of physics convenes every third year, in which, as a rule, the members of the Committee participate, and one year before the

[13]This fact has also been emphasised by Guido Bacciagaluppi and Anthony Valentini in their book entitled *Quantum Theory at the Crossroads: Reconsidering the 1927 Solvay Conference* [Bacciagaluppi and Valentini (2009)].

[14]This point will be discussed a little further on.

[15]Note from E. Verschaffelt to Lefébure, dated April 8, 1926. Archives of the Solvay Institutes, S.a.b.ULB, doc.2573. Verschaffelt was a member of the Administrative Commission who had taken part in the ISC meeting.

congress there is a preparatory meeting (of two days), as we just had. If a member cannot attend, that is no great misfortune, either. On the other hand, your cooperation would be of a very special value to us, in the first place because you have such a complete mastery of modern physics, but also for another reason. For, we have resolved, and this with the full concurrence of all the present members, as well as of the entire Administrative Commission (made up of Belgians), to invite Germans again next time. We are now hoping that, if you are a member of the Committee, this will contribute toward their accepting the invitation. We chose "The theory of quanta and the classical theories of radiation" as the topic for 1927, and we hope to have the following reports or presentations. You can see from this list that we have also placed hopes on you. Please regard this information as confidential for the time being:

1. W.L. Bragg: "New verifications of the classical theory".
2. A.H. Compton: "The Compton effect and its consequences".
3. C.T.R. Wilson: "Observation of photoelectrons and shock-electrons by the condensation method".
4. L. de Broglie: "Interferences and light quanta".
5. A note by Kramers "Theory of Slater–Bohr–Kramers and analogous theories".
6. Einstein: "New deductions of Planck's law and applications of statistics to quanta".
7. Heisenberg: "Adaptation of the foundations of dynamics to quantum theory".

It should be noticed that the term "conflict" does not appear in the theme of the Council evoked in this letter to Einstein. However, we will see that in a letter to Ehrenfest, dated January 18, 1927, Lorentz would cite the "conflict" between quantum theory and classical theories as a major topic of Solvay V.

On the other hand, there are some significant differences between the list of reports announced to Einstein, and the one transmitted two days later (i.e. on April 8) to Charles Lefébure, the Secretary of ISIP's Administrative Commission[16]:

[16]Note from E. Verschaffelt to Lefébure, dated April 8, 1926; Archives of the Solvay Institutes, S.a.b.ULB, doc.2573.

1. New verifications of the classical radiation theory (by L. W. Bragg).
2. The Compton effect and its consequences (by A. H. Compton or P. Debye).
3. Observation of photoelectrons and shock-electrons by the condensation method (by C. T. R. Wilson).
4. Interferences and light quanta (by L. de Broglie).
5. A note by Mr. Kramers on Slater's theory (if deemed useful by Kramers).
6. New derivations of Planck's law. Application of statistics to quantum theory (by Einstein or Ehrenfest).
7. Adaptation of the foundations of dynamics to quantum theory (by Heisenberg or Schrödinger).

Several factors may explain these differences. As for report 2, the idea of proposing Debye as an alternative to Compton seemed justified by the fact that the observation of a diminishing in frequency of light scattered by electrons had been made independently by Compton and Debye.[17]

Regarding report 6, the proposal "Einstein or Ehrenfest" was apparently made to Lefébure because of the uncertainty about Einstein's acceptance to present a report. It is also likely that the proposal "Heisenberg or Schrödinger" for report 7 was made because of the uncertainty about Heisenberg's reaction to the invitation. In case he would not come to Brussels, Schrödinger – a member of Solvay IV appreciated for his quantum interpretation of the Doppler effect – could report on the adaptation of classical dynamics to quantum theory.

On the other hand, there are elements that could explain the absence of second options in the list of reports communicated to Einstein. Lorentz had reasons to put forward arguments that might prompt his "touchy" colleague to reconsider the position he adopted in 1923, when he refused to take part in Solvay IV. Thus, by naming Heisenberg as the invited rapporteur of the last report, he may have wished to make clear that German fellows were not only invited to attend the Council, but were also expected to play a major role in it. With Einstein accepting the invitation, Heisenberg would almost surely accept to report on *matrix mechanics*. Finally, the option Debye in the letter to Lefébure may have been motivated by the high cost of transatlantic travels,[18] a matter of no concern to Einstein.

---

[17]See Compton's recollections of 1961 (section II.7.4).

[18]ISIP's rule was to limit reimbursement of travel expenses to Council members from Europe (see in the next section Lorentz's letter to Ehrenfest of January 18, 1927).

Yet, there is another way of explaining the differences between the two lists of reports. Regardless of the above considerations, it is possible that Lorentz started writing his letter to Einstein during the Committee meeting (i.e. on April 1 or 2, after obtaining the King's approval of the proposals made by the ISC), and that he kept part of his draft, with only one rapporteur for the reports 2, 6 and 7, when he finished the job on April 6. This possibility is suggested by the curious sentence in Lorentz's letter to Einstein: "The above-mentioned Committee, composed of Mrs. Curie, Langevin, Richardson, Guye (Geneva), Knudsen, van Aubel (Ghent) and me, is now in session."[19]

Whatever the reason for the gap between the two proposals, it is clear that by scheduling a presentation on the Compton effect immediately after Bragg's report on the successes of the classical radiation theory, the Committee decided that Solvay V would focus on the conflict between quantum and classical views on radiation.

Regarding Schrödinger, it is important to stress that when Lorentz informed Einstein of the reports to be discussed at Solvay V, he was (most probably) unaware of the new avenues that had been opened with Schrödinger's wave mechanics. [Schrödinger (1926a,b)] In fact, Schrödinger sent a letter to Lorentz ([Kox (2009)], doc 405) on March 30, i.e. two days before the ISC meeting, with a draft of two papers. One of them contained a proof of the equivalence between his wave mechanics and Heisenberg's "puzzling matrices."[20] Schrödinger also emphasised the fact that his approach was based on Louis de Broglie's ideas. However, it is likely that he sent his letter, and his papers, to Lorentz's address in Haarlem, unaware of the fact that Lorentz was in Brussels. Hence, we have reasons to assume that the news about Schrödinger did not reach Lorentz at the time of the ISC meeting, and that the idea of asking Schrödinger to report on "wave mechanics" came to Lorentz's mind only after Einstein's reaction to his letter of April 6, which contained this important news:

> As soon as the exclusion of Germans from the Solvay Institute, as a matter of principle, has ended, and Germans are invited again, both of which will be the case according to your letter, there is no reason for me to keep away from this friendly undertaking

[19]This statement is in contradiction with Lorentz's later words in that letter: "one year before the congress there is a preparatory meeting (of two days), as we just had."

[20]Schrödinger's paper [Schrödinger (1926c)], which contains his equivalence proof between matrix and wave mechanics, was received on March 18, 1926, and was still unpublished.

> any more. Therefore, I gladly accept my selection into the Committee. With regard to my own report at the 1927 Congress, with human foresight I could only tell what is already generally known. *Schrödinger*, however, has a theory of quantum states in press, a veritably ingenious implementation of de Broglie's idea. I would like to take the liberty of proposing he be given floor in my place [emphasis added].[21]

In addition, Lorentz must have been struck by Schrödinger's own words, when reading his letter of March 30: ([Kox (2009)], doc 405):

> I have now succeeded in clarifying the connection with Heisenberg–Born. It is such that all the "matrices" of the Heisenberg–Born theory can be calculated from wave mechanics by differentiation and quadratures, so that all the results of that theory with regard to the polarisation of the emitted light can be obtained from wave mechanics, and, I believe, can be given a much clearer interpretation ... The suggestion comes from the ingenious Thesis of Mr. L. de Broglie (*Annales de Physique* 10, n°3, 1925, 22) and the interesting remarks from Einstein (*Berliner Berichte.* 1925, 9).

Lorentz certainly needed more time to appreciate the progress made by Schrödinger. In fact, he wrote him a letter at the end of May ([Kox (2009)], doc 412), congratulating him for having shown that Heisenberg's matrices satisfy the equations of motion. Lorentz made clear that he preferred Schrödinger's approach to Heisenberg's, but asked for clarification on the use of configuration space. He kept exchanging letters with Schrödinger, asking for details on wave mechanics and its relation to de Broglie's ideas and Heisenberg's approach. Schrödinger, as we will see, was asked to present his wave mechanics after Einstein's refusal to report on quantum statistics.[22]

II.7.3 German participation to Solvay V: A tricky issue

Let us return for a moment to the ISC meeting of April 2, 1926. Briefed by Lorentz, who had met the King, the Committee decided to invite Planck and two other German physicists. These would be chosen among Born, Pauli and Heisenberg on the occasion of a later update of the Council's

[21] Einstein to Lorentz on April 12; ([Buchwald *et al.* (2021)], doc. 254).
[22] See [Kox (2009)], letters from Einstein (doc. 427) and to Schrödinger (doc. 428).

programme.[23] A noteworthy point was the absence of Sommerfeld's name on the list of invitees, while Bohr's appeared on top of that list. This absence, which could not be due to a simple oversight, was clearly shocking to the German invitees (the more so since Lorentz hold Sommerfeld in great esteem[24]). Born was extremely surprised to learn that the great Munich physicist had not been invited. He expressed his astonishment in a letter to Sommerfeld,[25] dated June 26, 1926. As for Planck, he informed Lorentz on June 13, 1926, that it was difficult for him to attend the Council, knowing that Sommerfeld, a man more familiar than him with the most recent advances in quantum theory, would not be present ([Kox (2009)], doc. 414):

> Dear Colleague,
>
> Your kind invitation reminds me of the first Solvay Congress in 1911, the theme of which was almost exactly the same as the one we are now planning again; and when I think of the harmonious and fruitful course of our conference at the time, I feel a particular pleasure at the honour bestowed on me by your kind invitation, and feel the desire to accept it if possible. But when I think of everything that has changed since that time, both in factual and personal terms, I have all sorts of doubts as to whether I am doing the right thing by travelling to Brussels, and I can show you my sincere gratitude no better than by trying to describe with the same sincerity what is on my mind with regard to this plan.
>
> Above all, unlike my feeling 15 years ago, I don't think that my presence would be of any significance to the work of the Congress. I am no longer one of those who are working at the forefront of the development of quantum theory, and who are called to participate in the first place ...
>
> The names of Heisenberg and Schrödinger, from whom the latest progress has come, give me particular satisfaction ... It will be very tempting later to compare the report of the 1927 Congress with that of 1911. What a contrast there will be! ... And now

[23]Pauli was born in Vienna but worked in Hamburg. The decision to invite Heisenberg had been taken before establishing the final list of Solvay V reports.

[24]On November 24, 1923, Lorentz had sent a letter to Sommerfeld, congratulating him for his "new great services to science" (see Haarlem, Noord-Hollands Archief, Archief H. A. Lorentz, (Archive 364)).

[25]See [Eckert and Märker (2004)].

> comes the concern that I must describe to you. If I, who am fully aware I am no longer necessary for successful work at the Congress, am nevertheless invited to participate, this could only be due to my previous involvement. If I now look at the list of participants at that time, I find a whole series of colleagues who are roughly in the same position as me ... I therefore cannot, and will by no means presume to criticise the list of invitees in any way. But, on the other hand, you will not blame me if I express the strong feeling that among earlier participants, whose invitation has been cancelled this time, there are personalities who are still working on the future development of quantum theory to a far more intensive degree than I am myself, and that this feeling is extremely depressing for me.
>
> I am thinking, among others, of my colleague Sommerfeld in Munich, who has actually devoted himself entirely to quantum theory since 1911, and whose significance for the development of the theory I need not describe to you ... Incidentally, there is still a long way to go until October 1927, and there is still a lot to think about. Under these circumstances, I would be doubly pleased to talk to you about all these things.

A point that Planck and Born were not aware of was that the Institute had "non-scientific" reasons for not inviting Sommerfeld. In 1926–1927, the members of ISIP's Administrative Commission could not accept the idea of welcoming a man who, during Belgium's occupation, had publicly supported Flemish activists (working at a dismantling of the country), and had taught German soldiers on military matters.[26]

II.7.4 More about the planned reports

With the exception of the last report (Heisenberg or Schrödinger), the focus at Solvay V was clearly on quantum theory and the problem of radiation. We already commented on the need to highlight the conflict between quantum and classical views on radiation, by scheduling a report on the "Compton effect" after Bragg's report. However, there is a decision with regard to the list of reports which deserves a few words of explanation:

[26]The authors thank Michael Eckert for providing them with this information. It is likely that Sommerfeld's Brussels lectures took place in Solvay's Institute of Sociology that had been requisitioned by the Germans.

the scheduling of a "Note" on "Slater's theory" (before withdrawing it from the final agenda).

At stake was the still puzzling problem of light quanta, and the possibility of a flaw in the principle of energy conservation, the strict validity of which had never been verified on the scale of atomic processes.[27] Two giants, Einstein and Bohr, were in deep disagreement on these subjects. As noted by Pais[28]:

> The first time that Bohr and Einstein found themselves in scientific opposition goes back to 1924, the final year of the old quantum theory ... At issue was their difference of opinion concerning the reality of light quanta.

In fact, Bohr's involvement in *Slaters theory*, an approach commonly known as the "unsuccessful BKS theory", may be explained, in part, by his persistent wish to avoid the introduction of 's light quanta. As mentioned by Ehrenfest in his letter to Lorentz of March 30, 1926, the BKS proposal [Bohr *et al.* (1924)], based on the assumption that energy-momentum is only conserved "statistically",was soon disproved by an experiment carried out by Walther Bothe and Hans Geiger [Bothe and Geiger (1924)].

A few years later, Heisenberg qualified the BKS idea as "the culmination of the crisis in the old quantum theory" ([Heisenberg (1929)], p. 490).[29] However, as noted by Heilbron ([Heilbron and Rovelli (2023)], p. 4), "the Bothe–Geiger experiment did not cancel the virtual oscillators (of BKS theory) or the conceptualisation of quantum mechanics as a calculus of observables". Furthermore, as pointed out by Jammer ([Jammer (1989)], pp. 188):

> The Bohr–Kramers–Slater paper was undoubtedly instrumental for the development of quantum mechanics in view of the fact that it made possible to establish a dispersion theory which offered a new approach to the fundamental problem concerning the interaction of matter and radiation.

[27]See [Pais (1991)], p. 233.

[28]See [Pais (1991)], p. 230 and our Appendix B for more details on Bohr's opposition to Einstein's light quantum hypothesis, the principle of energy conservation and the BKS proposal.

[29]See also [Pais (1991)], p. 236.

Indeed, the paper was "the point of departure of Kramers's detailed theory of dispersion, whose further elaboration by Kramers and Heisenberg, led the latter to the discovery of matrix mechanics, the first formulation of quantum mechanics" ([Jammer (1989)], p. 192).

Kramers obtained in 1924 a dispersion relation that described the emission probability of secondary radiation in terms of the incident radiation and Bohr's transition frequencies. He published two papers on the subject, and concluded ([Pais (1991)], p. 271):

> The obtained dispersion relation contains only such quantities – transition quantities referring to two stationary states – as allow of a direct physical interpretation on the basis of the quantum theory of spectra and atomic constitution, and exhibits no further reminiscence of the mathematical theory of multiple periodic systems with orbits.

Kramers's statement became, as indicated by Jammer, a source of inspiration for Heisenberg.[30]

The above elements provided a justification for scheduling a "Note" on *Slater's theory.* Its final withdrawal from the programme was probably due to Kramers's unwillingness to report on the subject. Yet, the BKS theory would be addressed, as we shall see, in Compton's report. Regarding this report, it should be stressed that Compton's interpretation of the *Compton effect* transformed Einstein's concept of a light quantum into a particle-like photon, endowed with an energy and a momentum. Given the importance of this step, it is worth recalling Compton's experiments on the absorption of scattered X-rays, and the support that Compton found in Maurice de Broglie's results on *corpuscular spectra*, as he struggled to get his results published in the *Bulletin of the National Research Council* [Compton (1922)]. The following lines are part of Compton's recollections [Compton (1961)]:

> From my absorption measurements I was able to estimate that over a wide range of wavelengths of the primary rays the increase in absorption of the scattered rays was what it should be if their wavelength was increased by about 0.03 Å over the wavelength of

[30] The difficulty with the concept of orbits had been highlighted by other physicists a year and a half before Heisenberg's breakthrough on Helgoland (see [Heilbron and Rovelli (2023)], p. 3).

the primary ray. This result I checked with an X-ray spectrometer, measuring an increase in the wavelength of approximately 0.02 Å.

At this point I found myself engaged, as a member of a committee of which William Duane of Harvard was the chairman, in preparing a report for the National Research Council on secondary radiations produced by X-rays. When it came to publication of the report, Duane objected to including my revolutionary conclusion that the wavelength of the rays was increased in the scattering process just described because he felt that the evidence was inconclusive. At the insistence of A. W. Hull, however, this portion of my report was included in the publication... At this point I paused in my experiments in order to concentrate on their theoretical interpretation ...

This led me to examine what would happen if each quantum of X-ray energy were concentrated in a single particle and would act as single unit on a single electron. Thus I was led to the now familiar hypothesis of an X-ray particle colliding with an electron and bounding classically from it with reduced energy, the lost energy appearing as the recoil energy of the electron. This idea of an X-ray quantum losing energy by collision with an electron must have been already in the mind of Peter Debye, then working in Zurich, for immediately upon the appearance of my report in the *Bulletin of the National Research Council*, he published a paper in the *Physikalische Zeitschrift*, in which he presented an explanation of the change in wavelength of the scattered rays, identical in principle with my own hypothesis, and appearing in print only a few days after my first publication ...

The results, confirming accurately the theoretical predictions, immediately became a subject of the most lively scientific controversy that I have ever known ... Having frequently repeated the experiments I entered the debate with confidence, but was pleased to find that I had support from P. A. Ross of Stanford and M. de Broglie from Paris, who had obtained photographic spectra similar to my own ... In the mean time other experimenters had not been idle. Within a few months after my first paper C. T. R. Wilson at Cambridge and Walter Bothe in Germany had found the recoil electrons predicted by the corpuscular theory ...

A. W. Simon and I, repeating these experiments, showed that the number of the tracks (left by the recoil electrons in the air

> traversed by X-rays) and their ranges, were just what should be expected according to the theory that each recoiling electron was the result of the impact of one photon of X-rays that it scattered.

As indicated by Compton, it was in 1925 that irrefutable evidence established the existence of the photon. Hence, in the early spring of 1926, the members of the ISC had every reason to schedule a presentation on the *Compton effect*, and to ask Wilson to report on "*photo- and recoil electrons*". As noted by Siegbahn,[31] the discovery of the recoil elections predicted by Compton was a double triumph. It established the validity of Compton's corpuscular radiation theory, and confirmed the importance and scope of Wilson's expansion method. Two months after Solvay V, these researchers shared the 1927 Nobel Prize for Physics.[32]

Let us return for a moment to the ISC idea of asking Heisenberg or Schrödinger to report on the "*Adaptation of the foundations of dynamics to quantum theory*". Today the proposal may seem strange, since the two approaches were totally different (see Dirac's speech at the University of New South Wales) as were Heisenberg's and Schrödinger's attitudes towards classical theory. Heisenberg came up in the summer of 1925 with a "quantum mechanics" involving only observable quantities. Together with Born and Jordan, he transformed it into an abstract formalism, called *matrix mechanics.* Schrödinger proposed a more conventional-looking "wave mechanics". His wish to conform to classical ideas has been emphasised by Pais ([Pais (1991)], p. 282):

> Odd as it may seem, it was Schrödinger's aspiration from the very beginning and forever thereafter, to incorporate, one might say, quantum theory into classical theory. Thus, in April 1926, he quoted Born and Jordan ([Born and Jordan (1925)], p. 858): "The new (matrix) mechanics presents itself as an essentially discontinuous theory", and then added: "Conversely wave mechanics, based on classical theory, represents a step towards a continuum theory".

As already mentioned, Schrödinger's work should have come as a great relief to Lorentz. The idea of asking Schrödinger to report on *wave*

[31] See Appendix D.

[32] The Prize was awarded on Siegbahn's proposal, this Swedish attendee of Solvay III having become a member of the Nobel Committee for Physics.

mechanics was put forward to him by Einstein in his letter of April 12, 1926. That it took Lorentz little time to share Einstein's enthusiasm is attested by his letter to Schrödinger of May 27 ([Kox (2009)], doc. 412):

> I specially liked the way in which you construct the appropriate "matrices", and show that they satisfy the equations of motion. This removes a concern that the works of Heisenberg, Born, Jordan, and Pauli had aroused in me.

On the other hand, it is likely that by the end of May, Lorentz had been informed of the works of Pauli and [Carl] Eckart, showing that *matrix mechanics* and *wave mechanics* are equivalent representations of the same state of affairs (Pauli announced his result to Jordan in a letter of April 12, 1926).[33] We will see that the idea of a Schrödinger report on "wave mechanics" materialised a few months later, on the occasion of an update of the Council's programme.

With regard to report 6, and the idea of discussing the quantum counterpart of statistical mechanics, it is worth recalling that in 1924 Satyendra Nath Bose had come up with a new derivation of Planck's law, based on the quantum statistics of indistinguishable light quanta.[34] This work made a great impression on Einstein, who, by extending Bose's ideas to matter, showed that at a sufficiently low temperature the atoms of a "Bose gas" would condense into a lowest energy state (a phenomenon called "*Bose–Einstein Condensation*"). Ehrenfest, was the first to whom Einstein communicated his idea of a condensation without attractive forces.[35] In the summer of 1925, he lectured in Göttingen and presented, among other developments, Louis de Broglie's prediction of *matter waves* ([Born (2005)], doc. 49).

On January 18, 1927, Lorentz sent Ehrenfest an invitation ([Kox (2018)], doc. 756):

> Before I go any further, I must write about the next Solvay meeting in Brussels. That will be in October (from 24 to 29), and I now have the pleasure to invite you on behalf of the Administrative Commission (Armand Solvay, Bordet, Bourquin, Henriot and Lefébure) and the Scientific Committee. Also invited are: Bohr,

[33]See [Pais (1991)], p. 203 note 94.
[34]See [Bose (1924)], p. 178.
[35]Einstein translated Bose's paper in German and sent it for publication to the journal *Zeitschrift für Phyisik* ([Bolles (2003)], p. 206).

Born, W. L. Bragg, L. Brillouin, L. de Broglie, A. H. Compton, Debye, [Henri-Alexandre] Deslandres, R. H. Fowler, Heisenberg, Kramers, Planck, Schrödinger and C. T. R. Wilson. The members of the Scientific Committee too (Mrs. Curie, van Aubel, W. H. Bragg, Einstein, C. E. Guye, Knudsen, Langevin, Richardson and myself) will be present at the meeting, and the physicists of the University of Brussels (De Donder, Henriot, and [Auguste] Piccard) will be invited. You already know the subject "Quantum theory and the classical theories of radiation". In particular we will discuss the conflict between old and new ideas, and attempt to arrive at a new mechanics. Verschaffelt will once again serve as secretary. We very much hope that you can come, and will be pleased if you do. Shortly, i.e. in the coming months, I will be able to send you further particulars of the reports etc. Later there will be an official invitation from the Administrative Commission, whose guests we shall be, and who will reimburse travel expenses (for visitors from Europe).

Chapter II.8

Update of the Council's agenda

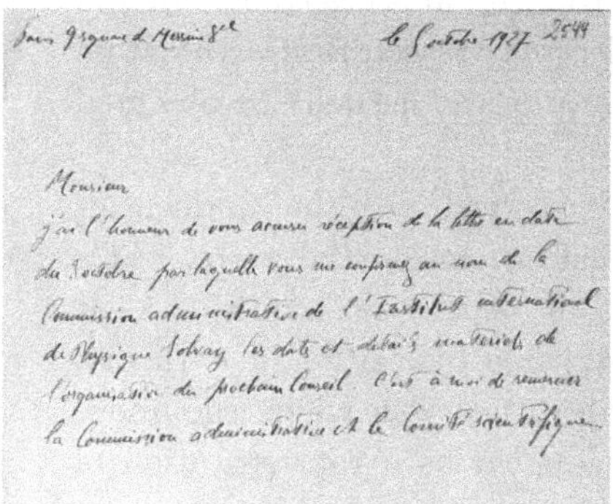

Paris 9 square de Messine 8e   le 5 octobre 1927   2544

Monsieur

J'ai l'honneur de vous accuser réception de la lettre en date du 3 octobre par laquelle vous me confirmez au nom de la Commission administrative de l'Institut international de Physique Solvay les dates et détails matériels de l'organisation du prochain Conseil. C'est à moi de remercier la Commission administrative et le Comité scientifique

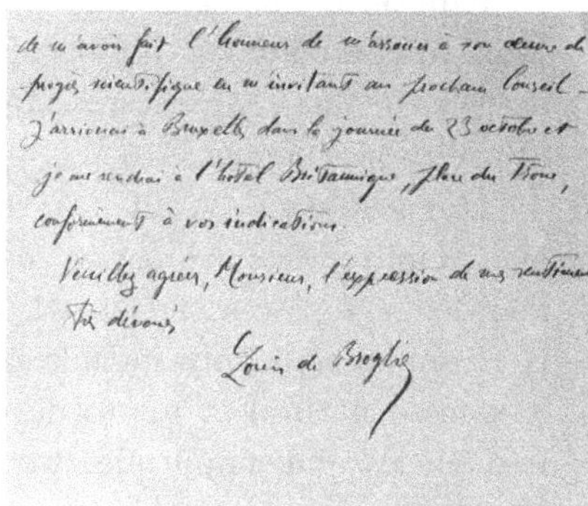

de m'avoir fait l'honneur de m'associer à son œuvre de progrès scientifique en m'invitant au prochain Conseil.

J'arriverai à Bruxelles dans la journée du 23 octobre et je me rendrai à l'hôtel Britannique, place du Trône, conformément à vos indications.

Veuillez agréer, Monsieur, l'expression de mes sentiments très dévoués,

Louis de Broglie

Louis de Broglie received an invitation to Solvay V in June 1927, and was asked to present a report. This letter contains details about his arrival in Brussels

Courtesy of the Solvay Institutes

Several crucial developments took place between January and October 1927. Heisenberg's idea of replacing classical dynamical variables by "infinite matrices", was followed in February 1927 by another feat: his formulation of uncertainty relations, providing the so-called "Copenhagen interpretation of quantum theory" with a quantitative expression of the fundamental wave-particle duality of light and matter.

II.8.1 Origin of Heisenberg's uncertainty relations

In the spring of 1926, Berlin had been the theatre of an important event: Heisenberg's first meeting with Einstein. Heisenberg had been invited to talk on "*Quantum Mechanics*" at the Berlin Physics Seminar.[1] After the talk, he was approached by Einstein, who wanted to discuss some delicate points. Here are Heisenberg's recollections of a crucial exchange that took place during the discussion ([Heisenberg (1972)], p. 94):

> **Einstein**: You agree that the atom contains electrons, but you want to eliminate the notion of electron trajectories in the atom despite the fact that electron trajectories can actually be observed in a Wilson chamber.
>
> **Heisenberg**: What can be observed are the frequencies and amplitudes of electronic oscillations inside the atom. This knowledge replaces the knowledge of "electronic orbits". As it is reasonable to only include observable quantities in a theory, it seemed natural to me to develop a theory in which the above frequencies and amplitudes would appear as a representation of the electronic orbits.
>
> **Einstein**: But you don't seriously believe that we should only introduce observable quantities into a physical theory... From a heuristic point of view, it can be useful to remember what you are really observing, but on the level of principles, it is completely wrong to want to base a theory solely on observable magnitudes. In reality things happen the other way round: only theory decides what can, and cannot be observed.

A few months later, Heisenberg was in Copenhagen, discussing the electron trajectory problem with Bohr. Here is Heisenberg's account of the

[1] We translated part of the exchange reproduced in the French version of Heisenberg's book [Heisenberg (1972)] (original title: *Der Teil und das Ganze*).

events that took place during the winter 1926–1927 ([Heisenberg (1972)], p. 112):

> Neither of us understood how a phenomenon as simple as the trajectory of an electron in a Wilson chamber could be reconciled with the mathematical formalisms of quantum mechanics or wave mechanics.
>
> In quantum mechanics the notion of trajectory did not exist. In wave mechanics a straight and oriented beam of matter could certainly exist, but it would gradually expand so as to cover spatial domains much larger than the diameter of an electron. The experimental situation, on the other hand, was undoubtedly different.
>
> In February 1927 Bohr decided to leave for a ski-holiday in Norway. I was quite happy to be able to reflect alone on this desperately difficult problem. An electron trajectory in a Wilson cloud chamber existed in reality since it could be observed. The mathematical formalism of quantum mechanics also existed. Hence there should be a way of establishing a link between one and the other. It was at that moment that I remembered Einstein's statement: "*Only theory decides what can, and cannot be observed.*"
>
> ... Bohr and I always admitted that it was possible to observe an electron trajectory in a Wilson chamber. But perhaps the only thing that we could see was a "discontinuous sequence of imprecise positions of the electron", each water droplet being much larger than an electron. Hence, the correct questions were the following:
>
> (i) Can we represent within the framework of quantum mechanics a situation in which an electron is approximately in a given position, i.e., up to a certain inaccuracy, and in which the electron possesses approximately a given velocity?
> (ii) Can we make the above imprecisions weak enough so as to avoid any contradiction with experiment?
>
> A brief calculation confirmed that such a situation could be represented mathematically, and that the inaccuracies were linked by relations, called later the uncertainty relations of quantum mechanics. These relations specified that the product of the uncertainties relative to the electron's position and to its momentum, could not be less than Planck's quantum of action.

Bohr on the other hand, continued to follow his thoughts. He tried to make wave-particle dualism the basis of his interpretation. The notion of "complementarity", introduced by him, was meant to describe a situation in which one and the same phenomenon could be grasped by two different (and contradictory) modes of interpretation. These modes were to be mutually exclusive and to complete each other. Only a juxtaposition of these contradictory modes would make it possible to fully exhaust the visual content of the phenomenon.

Our next priority was to present the situation, now fully understood by us, in a way that would make it understandable to our colleagues. Two opportunities to confront other physicists presented themselves in the autumn of 1927. One was a general physics congress organised in Como, where Bohr gave a detailed report on the new situation in quantum mechanics. *The other was the Brussels Solvay congress, where, according to the Foundation's habits, only a small number of specialists were invited to discuss the problems of quantum theory in full detail.* [emphasis added]

II.8.2 Update of the list of participants to Solvay V

As mentioned by Heisenberg, the fifth Physics Council was to take place during the last week of October 1927. Several months earlier, as Lorentz returned from his third trip to the United States, he was informed of the latest developments regarding the organisation of Solvay V:

- Einstein, who had been asked to talk about his work on perfect gases, had made clear that he did not wish to present a report.
- W. H. Bragg, a British member of the ISC, had announced that he would not be able to attend the Council, and that he wished to retire from ISIP's Scientific Committee.
- van Aubel, the Belgian Committee member, had informed his colleagues of his refusal to take part in a conference that would be attended by German physicists.

Disappointed by the news, Lorentz set out to find new Council members. He consulted Ehrenfest again on the most recent advances in quantum theory.[2] Possible invitees could be chosen among those suggested by

[2]See also section II.7.2.

Ehrenfest for spending some time in Leiden during the summer of 1927. Names of possible guests were mentioned by Ehrenfest in the following letter to Lorentz,[3] dated June 14, 1927:

> Dear Mr. Lorentz,
>
> The names of physicists that I propose for a stay in Leiden are the following:
>
> - Born, Schrödinger: for visits of one or two weeks, so that we can get an overview from them in almost daily lectures "on how to see the latest developments". Heisenberg, Pauli: for two or three weeks (a longer stay might be impossible for them).
> - Jordan: is unfortunately impossible because he stutters and it is a torture to listen to him.
> - [Walter] Gordon: has given a particularly clear presentation of the Compton effect. I don't know him personally. We could learn a lot from him if he stayed for 4 to 6 weeks.
> - [Fritz] London: I don't know him personally. He seems to be somewhat mathematical and formal, so he could learn something from us. I would have a lot to learn from him. Goudsmit knows him well and appreciates him. I would like to have him here for 1 or 2 months.
> - Francis Perrin: I could learn a lot from him, but not what I want to learn most of all now: the Dirac–Heisenberg–Klein–Hund area. Fraulein [Lucy] Mensing would help the younger ones to get to grips with the newer things through her presence.[4]
> - [Charles] Manneback (Belgium): Has worked in Zurich with Debye. Edmond Bauer told me about him with great warmth. He would like to study in Leiden for some time.
> - Fowler (Cambridge!).
> - [Llewellyn] Thomas (Cambridge): Is unfortunately almost completely incapable of making his thoughts clear to us.

[3]The authors thank Frits Berends for providing them with a copy of Ehrenfest's letter to Lorentz of June 14, 1927.

[4]According to Pasual Jordan's recollections, Lucy Mensing was asked by Jordan to investigate one of the fist applications of matrix mechanics to diatomic molecules. See the transcript of Pascual Jordan's Interview – p. 50 – conducted by Thomas S. Kuhn on 17, 18, 19, and 20 June 1963. `https://repository.aip.org/jordan-ernst-pascual-1963-june-17-18-19-and-20`.

- [Robert] Oppenheimer: Is a particularly fine and talented young American, who was present at the American week in Leiden. Has now worked in Cambridge, Copenhagen and Göttingen (where he still is). Writes terribly badly. I could learn a lot from him. But with his way of'dealing with problems, a long-term collaboration would hardly be possible.
- [Gregor] Wentzel (Professor in Leipzig): Is a student of Sommerfeld.

NB: It would be much more productive for us to have two guests at the same time (if, of course, they are compatible with each other) than one after the other. For example [Igor] Tamm and Gordon (if Gordon is not totally different from what I am told). It would be very desirable for me to have my old high school friend, Prof. [Gustav] Herglotz (Göttingen), in Leiden as a mathematical guest for 7 or 10 days (similar to [Richard] Courant's visit), sooner or later, as he has a wonderful ability of extracting things from mathematics that can inspire physicists (Because of his mother's serious illness I cannot make any specific plan for the moment).

I would also like to ensure that we are all well informed about the very important and wonderful investigations in the Göttingen laboratory through the collaboration of Franck (perhaps [Otto] Oldenburg, who was in California for a longer period of time). Despite repeated discussions at the Colloquium, we are very poorly informed. Our overview of Stern's work is also poor.

We also know from Ehrenfest's letter that Lorentz wished to be kept informed of any recent result likely to enrich the Council debates. These are the three publications, communicated by Ehrenfest, on topics he had discussed with Lorentz:

(i) Elsasser, W. [Walter] (1925). Bemerkungen zur Quantenmechanik freier Elektronen,[5] *Naturwissenschaften* **13**, p. 558.
(ii) Dymond, E. G. [Edmund] (1926). Scattering of Electrons in Helium, *Nature* **118**, p. 336.
(iii) Davisson, C. [Clinton] and Germer, L. H. [Lester] (1927) The Scattering of Electrons by a Single Crystal of Nickel, *Nature* **119**, p. 558.

[5]In English: "Remarks on the quantum mechanics of free electrons".

As we will see, the wealth of information on recent achievements, patiently transmitted by Ehrenfest to Lorentz, proved decisive for the Council's success.

Meanwhile, Louis de Broglie had continued to elaborate on his ideas. As he recalled many years later ([L. de Broglie (1947)], p. 138):

> [To] really remain in agreement with classical ideas, it was necessary to succeed in incorporating the corpuscle into the wave in such a way as to obtain a clear and coherent image of the wave-particle dualism in the framework of space and time.

It was in this spirit that Louis created his "double-solution" approach.[6]

In June 1927, Lorentz wrote a letter to Louis de Broglie, asking him to report on "Wave mechanics and its interpretation". However, having realised that, apart from some simple situations, his "double-solution" approach involved enormous mathematical difficulties, Louis thought of establishing himself first more firmly on a "fallback position, less satisfactory in principle, but perhaps easier to defend."[7] In this attempt, nowadays known as *the pilot wave approach*, classical point-like corpuscles were assumed to be driven by continuous waves "transposed from physical optics to wave mechanics" ([L. de Broglie (1947)] p. 185).[8]

Regarding possible new Council members, Lorentz named two physicists in a message to Lefébure,[9] dated August 27:

> Since last year, quantum mechanics, which will be our subject, has developed with unexpected rapidity, and physicists who first came in second position, have made extremely remarkable contributions. That is why I would be happy to invite Mr. Dirac from Cambridge, and Mr. Pauli from Copenhagen.

Lefébure agreed. As a result, the members of Solvay V benefited from the presence of ten major contributors to the quantum revolution: Bohr Born, Compton, Debye, de Broglie, Pauli, Heisenberg, Schrödinger, Dirac and C. T. R. Wilson. The updated list of invitees and that of the invited reports finally looked as follows.

---

[6]For further details, see [Bacciagaluppi and Valentini (2009)].

[7]See [L. de Broglie (1947)], p. 185.

[8]See [Bacciagaluppi and Valentini (2009)] for a comparison between the *double-solution* and *pilot-wave* approach.

[9]Archives of the Solvay Institutes, S.a.b.ULB, doc.2523.

Updated list of invitees

Chairman: H. A. Lorentz; *members*: M. Curie, N. Bohr M. Born, W. L. Bragg, L. Brillouin, A. H. Compton, P. Debye, L. de Broglie, P. A. M. Dirac, P. Ehrenfest, A. Einstein, R. H. Fowler, Ch.E. Guye, W. Heisenberg, M. Knudsen, H. A. Kramers, P. Langevin, W. Pauli, M. Planck, O. W. Richardson, E. Schrödinger, C.T.R. Wilson; secretary: E. Verschaffelt.

Updated list of invited reports

1. *X-ray reflection intensities*, by W. L. Bragg.
2. *Discrepancies between experiment and the electromagnetic theory of radiation*, by A. H. Compton.
3. *The new dynamics of quanta*, by L. de Broglie.
4. *Quantum Mechanics* by M. Born or W. Heisenberg.
5. *Wave Mechanics* by E. Schrödinger.[10]

In fact, the Minutes of Solvay V, published by Gauthier-Villars, contain a sixth report by Bohr, entitled "*The Quantum Postulate and the New Development of Atomistics.*" Bohr as we know, was not expected to present a report. However, Lorentz decided to include his "Como presentation" in the Gauthier-Villars volume and to present it as a Solvay report ([Solvay V (1928)]).[11]

II.8.3 A last-minute guest

Solvay V was attended by the American physical chemist Irving Langmuir. Eager to invite this well-known scholar who happened to be in Europe, Lorentz sent a telegram to Lefébure on October 19 (i.e. five days before the meeting) with the following message: "Langmuir, excellent American physicist now in Paris. May I invite him to the Council?"

Lefébure reacted positively. Langmuir, on the other hand, was intrigued, as witnessed by this message to his mother ([Coffrey (2008)], p. 188): "This year, the Council focuses on quantum theory, an area to which I have not contributed at all. So, I do not see why I am invited."

[10] As already noted by Bacciagaluppi and Valentini, due to a typo, Schrödinger's name is missing from the published list of invited members.

[11] See also the comments made by Bacciagaluppi and Valentini about this subject ([Bacciagaluppi and Valentini (2009)], p. 243).

Langmuir's last-minute invitation deserves an explanation. This future American Nobel laureate in chemistry was known for his work in atomic physics. His ideas on the distribution of outer atomic electrons had been praised by Rutherford at Solvay III: "Langmuir's theory – he said – is purely descriptive and qualitative, but is undoubtedly useful because it indicates the possible arrangements of electrons at the atom's periphery, and accounts for some of the principal physical and chemical properties of the atom."

But this fact alone did not justify Langmuir's presence at a meeting that had been convened to discuss classical and quantum properties of radiation, and to react to the most recent quantum theoretical proposals. His invitation was an operation orchestrated by Léon Brillouin, who wanted to get the American chemist to Brussels. Brillouin's motives are explained in Patrick Coffey's book ([Coffrey (2008)], pp. 188–190):

> Léon Brillouin, the French physicist and an invitee to the conference had been in contact with Langmuir's boss Willis Whitney, trying to negotiate a consulting contract with General Electric, and Whitney had asked Brillouin to connect with Langmuir at an earlier conference in Como, so he apparently fudged things a bit to get Langmuir an invitation to the Solvay Conference. He led Lorentz to believe that Langmuir was going to be in Brussels anyway that week, and asked him to extend Langmuir an invitation. Langmuir jumped at the chance and later sent Whitney a report of the conference.

This tells us that Lorentz acted at Brillouin's request. But why did this picky chairman, who refused to admit auditors to the Councils, suddenly agree to break[12] this rule? His decision to accept Brillouin's request, was linked to an unexpected event: the celebration by the French Physical Society of the Augustin Fresnel centenary.[13] The members of Solvay V had been invited to attend the ceremony scheduled on October 27 – in

[12]Lorentz already bent his rule by accepting the presence at Solvay V of three physics professors of the Université Libre de Bruxelles (Piccard, de Donder, and Jacques Errera). He explained that it would be shocking for them to meet ISIP's guests at the University reception and at Mr. Solvay's dinner, and not be invited to attend the Council sessions. It is also worth mentioning that the possibility of inviting American physicists passing through Europe was also mentioned (and informally accepted) at the ISC meeting dated March 30, 1920. See Verchschaffelt's notes L11/52_bis ESPCI.

[13]Details of the event can be found in [Lambert and Berends (2021)], p. 231.

the middle of the Council – at the Paris Sorbonne. This invitation was a source of embarrassment, especially for Lorentz, who had been invited to deliver a speech in Paris. Also Louis de Broglie and Pieter Zeeman had been invited to make presentations. In short, Lorentz needed Brillouin's help to resolve urgent questions regarding the event. He decided that the Council sessions scheduled for the afternoon of Thursday 27 would be suspended, and that the debates would resume on Friday 28 (with a possible extension to Saturday 29).

Langmuir's presence in Brussels proved beneficial for various reasons. The letter to his mother provides us with a vivid account of the Council sessions and the general feeling at the end of the conference:

> Strenuous meetings – 8 hours per day in a lecture room and all evenings (often till one o'clock in the morning) in the hotel where we all stayed discussing quantum theory. The wave mechanics is the universal mode of expression today. For the first time quantum theory can be formulated in apparently complete form, so that the discrepancy between the classical wave theory and quantum phenomena (even the photon theory) seems to be clearing up.

Another bonus was the fact that the American chemist had a camera, and used it to film the Council members walking on Brussels's Grand-Place. The rare footage is a *home movie* shot with a *General Electric* camera. The movie captures 2 minutes of an intermission in the proceedings. Twenty-one of the 29 attendees appear in the film.[14]

[14]A link to the footage can be found in the *Controversies* section of the *The Solvay Science Project* virtual exhibit. For the English version of the website and the link to Langmuir's footage, please visit `http://www.thesolvayscienceproject.be/exhibits/show/the-solvay-science-project/controverses--th--me-7-/solvay-v--1927`.

Chapter II.9

Solvay V (1927): Electrons and photons

Solvay V

Courtesy of the Solvay Institutes

II.9.1 Announcement of Solvay V

The fifth Physics Council took place, as expected, from October 24 to 29. A few days before the meeting Lorentz sent a Note to the Brussels newspaper *Le Soir*, highlighting the connection between the programme of the meeting and the discussions that had taken place at Solvay I:

> The issues dealt with by the present Council are closely linked to those of sixteen years ago. Mr. Planck had introduced the notion of discontinuities and abrupt transitions in the motion of atoms and electrons, and in 1911 the question was to appreciate the role that these discontinuities were playing in various phenomena, and examine the fundamental laws that govern them.
>
> The first exploration of a new and vast terrain, the richness of which had been revealed during the years that have passed since then, already suggested the need to reform the foundations of mechanics, so as to account for the discontinuities in question (their magnitude is now referred to as "quanta") and characterise them, not as something additional but as a fundamental and essential element.
>
> This year the discussions will focus on the attempts that have been made in recent years to develop what may be called "quantum mechanics", and in which Messrs. de Broglie, Heisenberg, Born, Schrödinger, Dirac, and others took part. The work of these physicists still constitutes a set of a more or less disparate aspect, because in spite of the unity which exists deep inside, great divergences of opinion remain. As a result, the "clash of opinions" will not be lacking. It is in such circumstances that the "Solvay method" is particularly efficient, and that it has the ability to clarify ideas and to accelerate progress.

Lorentz, was right. The fifth Physics Council was characterised by contrasting factors. On the one hand, it marked a reconciliation of physicists who had been separated by the war. On the other, it became the theatre of a "clash" between two generations: the "*boys' club*", represented by Pauli, Heisenberg and Dirac, versus the "*old guard*" which included Planck, Lorentz, Einstein and Schrödinger.

II.9.2 Highlights of the fifth Physics Council

As mentioned by Heisenberg, and recalled by Coffey, Solvay V was preceded by the "Volta conference", held in Como from 11 to 20 September to commemorate the centenary of the death of Alessandro Volta. Einstein, who was not present in Como, was eagerly awaited in Brussels. Everyone wanted to know how he would react to a "revolutionary mechanics", based as Heisenberg put it ([Heisenberg (1989)], p. 47), "on a mathematical formalism which made impossible a simultaneous and perfectly precise description of the position and the velocity of an electron." In fact, Einstein remained extremely discrete. He intervened just once, analysing a thought experiment which led him to conclude that "quantum mechanics" could not be considered a complete theory.

A detailed analysis of the reports and discussions at Solvay V can be found in the book published in 2009 by Bacciagaluppi and Valentini (to which we referred in section II.7.2), which contains an English translation of the Council *Minutes* [Bacciagaluppi and Valentini (2009)].

Here, we limit ourselves, once more, to reporting elements that shed light on our point. Three topics were at the centre of the debates: the *duality* of light, confirmed by the *Compton effect*; the *duality* of matter, predicted by Louis de Broglie and underpinned by the recent discovery of matter waves[1]; the fundamental indeterminacy at atomic scale, described by Heisenberg's *uncertainty relations*, and generalised by Bohr's *complementarity principle.*

In what follows, we focus on some noteworthy points that are part of the Minutes published by Gauthier-Villars in 1928. We also mention an illuminating conversation between Bohr and Dirac, that was reported by Dirac in his Lectures at the University of New South Wales.

II.9.2.1 *The conflicting aspects of light*

Bragg's analysis of accumulated X-ray data showed that the high precision achieved by Siegbahn, and other experts in the field of X-ray spectroscopy, was "the best proof of the accuracy of the classical laws of optics". Having heard Bragg's arguments in favour of the classical electromagnetic theory,

[1] It is worth emphasising that the experimental results, obtained by Davisson and Germer, had been preceded by an analysis made in 1926 by Elsasser. His conclusions, published in *Naturwissenschaften*, were mentioned by Born.

the participants were presented evidence in support of Compton's photon theory, based on the observation of "directed" light quanta.

In the first part of his report, Compton listed the difficulties encountered by classical views, according to which radiation consists of electromagnetic waves propagating through space in accordance with Maxwell's equations. He mentioned Bohr–Kramers–Slater proposal, or the *virtual radiation theory*, which aimed at reconciling wave theory with the existence of stationary atomic states.

Arthur Compton in 1927
Courtesy of the Nobel Foundation archive

Kramers and Einstein in 1927
Courtesy of the Solvay Institutes

This approach, based on the assumption that excited atoms continually emit virtual radiation, implied a renunciation of energy conservation in its strict acceptance (the energy of emitting or absorbing atoms being only conserved "statistically"). Compton also pointed at the difficulty for wave theories to avoid the embarrassing concept of the ether: "If we abandon the idea of an ether, it is simpler to assume that the energy of radiation is transmitted in the form of corpuscles rather than in the form of waves." Reporting on photoelectric effects, Compton praised "M. de Broglie's beautiful experiments with the magnetic spectrograph" because they showed that:

Einstein's photoelectric equation could be applied to the region of X-ray frequencies, on condition of interpreting the term with index "p" as the work necessary to extract the electron from the energy level "p" in the atom.

At this point, Compton referred to the experiment of Joffé and Dobronravoff [Joffé and (1925)] – presented at Solvay IV and published soon after the Council – showing that the idea of an "accumulation" of the energy transferred to the expelled electron was extremely hard to defend. Turning his attention to *photon theory*, Compton argued that the theory predicted, quantitatively and in all its details, the change in wavelength of the scattered X-rays and the properties of the recoil electrons.[2] He admitted that these results, in contradiction with classical theory, could be explained by the "*virtual radiation theory*", but stressed the incapacity of this theory to predict them. Having pointed at the Bothe–Geiger experiment [Bothe and Geiger (1924)], which allowed to decide between "photon theory" and "virtual radiation theory", Compton ended his report with an impressive list of arguments in favour of the former[3]:

> The classical theory that radiation consists of electromagnetic waves propagated through space in all directions is intimately connected to the idea of the ether, which is difficult to conceive. It affords no adequate picture of the manner in which radiation is emitted or absorbed. It is inconsistent with the experiments on the photo-electric effect, and is entirely helpless to account for the change in wavelength of scattered radiation or the production of recoil electrons.
>
> The theory of virtual oscillators and virtual radiation, which is associated statistically with sudden jumps of atomic energy and emission of photo-electrons and recoil electrons, does not seem to be inconsistent with any of these phenomena, as viewed on a macroscopic scale. This theory, however, retains the difficulties inherent in the conception of the ether, and seems powerless to predict the characteristics of photo-electrons and recoil electrons. It is further difficult to reconcile with the composite

[2]Compton recalled the discovery of the recoil electrons a few months after their prediction, independently by Wilson and Bothe.

[3]Compton's statement is also reported in [Bacciagaluppi and Valentini (2009)], p. 354.

> photo-electric effect, and is also contrary to Bothe's and Geiger's coincidence experiments and to the ray track experiments relating the directions of ejection of a recoil electron and of emission of the associated scattered X-ray.
>
> The photon theory avoids the difficulties associated with the conception of the ether. The production and absorption of radiation is very simply connected to the modern idea of stationary states. It supplies a straightforward explanation of the major characteristics of the photo-electric effect, and it accounts in the simplest possible way for the change in wavelength accompanying scattering and the existence of recoil electrons. Moreover, it predicts accurately the results of experiments with individual radiation quanta, where statistical theory fails. Unless the four experiments on individual events are subject to improbably large experimental errors, the conclusion is, I believe, unescapable that radiation consists of directed energy quanta, i.e. photons, and that energy and momentum are conserved when these photons interact with electrons and atoms.
>
> Let me say again that this result does not mean that there is no truth in the concept of waves of radiation. The conclusion is rather that energy is not transmitted by such waves. The power of the wave concept is too well known to require emphasis. Whether the waves serve to guide the photons, or whether there is some other relation between photons and wave is another and difficult question.

Reacting to Compton's report, Lorentz made an ultimate plea in favour of the electromagnetic ether[4]:

> Mr. Compton considers an advantage of the photon theory that it allows us to do without the hypothesis of an ether, which entails great difficulties ... In my opinion, the theory of relativity does not necessarily exclude the notion of a universal medium ... Secondly, it is quite certain that, in luminous phenomena, there must be something other than photons ... There must be something which guides them in their motion, and it is natural to associate that something with the electromagnetic field, as determined by the

[4]The concept of the ether survived the acceptance of Einstein's relativity, but its role changed. For an in-depth analysis, see [Navarro (Ed.)].

classical theory. This notion of an electromagnetic field, with its waves and vibrations, would bring us, according to Mr. Compton's way of seeing, to the conception of an ether.

Various members, notably Bragg, Wilson and Richardson, expressed doubts about the validity of Compton's law, according to which the average component of the momentum of the electrons in the direction of the photon's motion should be equal to the momentum of the photon, whose energy has been absorbed and is carried by the photo-electron. Bohr, for his part, made this general remark on the problem of quantum theory:

> The dilemma we face regarding the nature of light is only a typical example of the difficulties we encounter when we want to interpret atomic phenomena by means of classical concepts ... In accordance with the fundamental ideas of Mr. de Broglie which found such perfect confirmation in the experiments of Davisson and Germer, the notion of waves is just as essential in the interpretation of material particles as in the case of light. We know from this that it is also necessary to attribute to the wave field a finite extent in space and time, if we want to define the energy and momentum of an electron, just as in the case of the quantum of light we must admit such finite extent so that we can speak of frequency and wavelength.

II.9.2.2 *Louis de Broglie on Schrödinger and Born*

Louis de Broglie began his report with a reminder of his early work, which led to a simple interpretation of the quantum theoretical Bohr–Sommerfeld conditions. He recalled that the application of his concepts to the corpuscles of light allowed him to account for the Doppler effects, the radiation pressure, the photoelectric effects and the Compton effect. He then gave an account of Schrödinger's fundamental idea that the atom should be considered as the seat of standing waves that satisfy specific conditions, leading to a set of *eigenfunctions* which represent the stable states of the atomic system.

De Broglie emphasised Schrödinger's "*admirable ingenuity*", which enabled him to show that the new mechanics, invented by Heisenberg and developed by Born, Jordan, Pauli and others, could be translated into the language of wave mechanics. He nevertheless pointed at the difficulties raised by Schrödinger's conception of an electron "dissolved into the atom" so that one could no longer speak of its position or velocity. As for the ideas developed by Born, de Broglie saw no reason to renounce determinism in individual physical processes, arguing that Schrödinger's wave function could be seen as describing a "*pilot wave*" (called *Führungsfeld* by Born), but also as a "*probability wave*". Hence, he concluded: "It is in this that our conceptions, very similar to those of Mr. Born, nevertheless differ significantly from his."

II.9.2.3 *Heisenberg on quantum mechanics*

Heisenberg started his report by recalling the empirical basis of *matrix mechanics*: Walther Ritz's combination principle, according to which the frequency ν of each spectral line of an atom can be written as the difference of two terms, carrying both a numerical index. This meant that the set of all lines of an atom could be best described by means of a square-shaped "diagram", with rows and columns, i.e. by a matrix.[5] The Born–Heisenberg report provided the Council members with the following description of the new mechanics:

> Quantum mechanics is based on the idea that atomic physics is essentially distinguished from classical physics by the existence of discontinuities. Quantum mechanics should be seen as a direct extension of quantum theory, established by Planck, Einstein and Bohr. Already before the birth of quantum mechanics, Bohr often insisted on the fact that these discontinuities should lead to the introduction of new notions in kinematics and mechanics, so that in any case classical mechanics, and the system of concepts corresponding to it, should be abandoned. Quantum mechanics seeks to introduce new notions through a precise analysis of what is "essentially observable" ...
>
> The new system of concepts gives, at the same time, the intuitive content of the new theory. From an intuitive theory we must only ask that it be in itself without contradiction, and that it leads to an unambiguous prediction of the possible results of any imaginable experiment in its domain. It is in this sense that Quantum Mechanics should be regarded as an intuitive and complete theory of micro mechanical processes.

---

[5]See Dirac on the origin of Heisenberg's matrices. Section II.6.2)

In his presentation of the report, Heisenberg first focused on the mathematical methods of quantum mechanics. He then turned to its physical interpretation, the formulation of its basic principles, and the delimitation of their scope, giving due attention to the role played by Born, Jordan, Dirac and Schrödinger. His report shed new light on the physical significance of Planck's quantum of action. "The true meaning of Planck's constant is therefore this, that it constitutes the universal measure of indetermination which is introduced into natural laws by the wave-particle dualism."

Heisenberg also addressed the current applications of quantum mechanics. His conclusion left no doubt about the views he shared with Born on the "definitive" character of quantum mechanics:

> In summary, we would like to point out that we do not consider the research mentioned last, which is related to the quantum mechanical treatment of the electromagnetic field, as finished, but that we consider quantum mechanics to be a complete theory, whose fundamental physical and mathematical hypotheses are no longer susceptible of modification ...
>
> The dualism between corpuscles and waves, which in quantum theory appears as part of a complete theory, without contradictions, exists in a completely similar way for radiation. The relationship between light quanta and electromagnetic waves must be just as "statistical" as the one between de Broglie waves and electrons.

II.9.2.4 *Schrödinger on wave mechanics*

As emphasised in the Born–Heisenberg report, Schrödinger recalled that the equivalence between *matrix mechanics* and *wave mechanics* marked the birth of a an all-encompassing "quantum mechanics". However, he insisted on the distinction that should be made between a four-dimensional mechanics, and the multidimensional wave mechanics that provides the mathematical solution to the problems posed by the Heisenberg–Born matrix mechanics.

Schrödinger nevertheless expressed the hope of being able to understand the results of this multidimensional mechanics in the customary three-dimensional space. His statement gave gave rise to a scathing reaction from Heisenberg: "I see nothing in Mr. Schrödinger's calculations which justifies his hope". To Bohr, who intervened by asking "Hasn't Dirac found the solution to this problem?", Schrödinger replied:

> Dirac's results are certainly very interesting and point the way towards a solution, if they do not already contain it. Only we must first understand each other in physical language. I do not yet see an answer to a physical question in the assertion that certain quantities are subject to a non-commutative algebra, especially when it comes to quantities which must represent numbers of atoms.

II.9.2.5 *Dirac: Quantum theory versus relativity*

In his New South Wales Lectures, Dirac recalled a conversation he had forty years earlier with Bohr:

> I remember, in particular, an incident in the Solvay conference of 1927. During the interval before one of the lectures, Bohr came up to me and asked me "What are you working on now?" I tried to explain him that I was working on the problem of trying to find a satisfactory relativistic quantum theory of the electron. And then Bohr answered that that problem had already been solved by Klein.
>
> I tried to explain to Bohr that I was not satisfied with the solution of Klein, and I wanted to give him reasons, but I was not able to do so because the lecture started just then, and our discussion was cut short. But it rather opened my eyes to the fact that so many physicists were quite complacent with a theory which involved a radical departure from some of the basic laws of quantum mechanics, and they did not feel the necessity of keeping these basic laws in the way I felt.
>
> Well, I was worrying over this point for some months, and then, ultimately, I found another wave equation where we now have a "psi" involving four components instead of just a single "psi".

It seems undeniable that this short conversation with Bohr may have convinced Dirac that no one else was concerned with the need to obtain an "acceptable" relativistic equation for the electron. This fact may have encouraged him to continue an effort that led, a few months later, to a resounding success: Dirac's relativistic equation.[6]

[6] Dirac's success is mentioned by Bohr in his post-Council report added by Lorentz to the official proceedings. See [Solvay V (1928)], p. 246.

II.9.2.6 *Einstein on quantum mechanics*

Einstein intervened, as we already said, only once during the general discussion, applying two possible conceptions to a simple situation involving electrons:

> **Conception I:** The de Broglie–Schrödinger wave does *not* correspond to a single electron, but to a cloud of electrons extended throughout space. The theory gives no information on individual processes, but only on the whole of an infinite number of elementary processes.
> **Conception II:** The theory claims to be a complete theory of individual processes, each particle being described by a de Broglie–Schrödinger wave packet.

Einstein (1879–1955) at Solvay V
Courtesy of the Solvay Institutes

According to Einstein, it was only by virtue of conception II that the theory implied the validity of the conservation laws for elementary processes. It was also from conception II that the theory could account for the result of the Geiger-Bothe experiment, and that it could explain the fact that in a Wilson cloud chamber the droplets corresponding to an alpha-particle are "approximately" situated on a continuous line.

Despite these facts, Einstein raised objections to conception II, which, in his opinion, could only be lifted by assuming that the process can be described by a Schrödinger wave, and also that the particle can be localised during the propagation of the wave.

As for de Broglie's approach, Einstein declared: "I think that Mr. de Broglie is right to look in this direction. If we operate only with Schrödinger's waves, the interpretation II implies, in my opinion, a contradiction with the postulate of relativity."

As pointed out by Bacciagaluppi and Valentini ([Bacciagaluppi and Valentini (2009)], p. 175), Einstein's contribution to the general discussion contained a "seed" of what would become known as the *EPR argument*, developed in the Einstein–Podolsky–Rosen paper of 1935 [Einstein *et al.* (1935)].[7] It is in this sense that one may argue that Solvay V contributed, to some extent, to the emergence of a *second* quantum revolution.[8]

II.9.3 Louis de Broglie's recollections of Solvay V

Recalling many years later the discussions at the fifth Physics Council, Louis de Broglie gave this brief account of the different attitudes towards wave-particle dualism ([L. de Broglie (1976)], p. 15):

> Though the concepts introduced by wave mechanics were generally accepted at the time, the details of the physical interpretation of wave-particle dualism remained very controversial ... Schrödinger and myself tried to present a concrete and causal interpretation, by and large in agreement with the traditional concepts of physics ...
>
> Born, Bohr, Heisenberg, Pauli, and Dirac, presented a novel, purely probabilistic interpretation, based on Heisenberg's recently discovered uncertainty relations.

De Broglie recalled that the last group of physicists, representing the indeterministic school, met his theory with cold disapproval. Lorentz, on the other hand, "spoke out as a convinced partisan of determinism and of the concrete images provided by classical physics". De Broglie also emphasised Einstein's positive appraisal of his work: "Although he never expressed the opinion that I had found a definitive solution to the problem, he always encouraged me in private to persevere on my path." The clearest

[7] Boris Podolsky and Nathan Rosen coauthored the paper with Einstein.

[8] The EPR argument stimulated John Bell in his investigation of situations with two "entangled" particles (See [Bell and Aspect (2004)]).

sign of sincere approval was the one Einstein gave him on the occasion of the Fresnel celebration at the Sorbonne ([L. de Broglie (1976)], p. 16):

> On the return trip from Brussels to Paris ... I had a final conversation with Einstein on the arrival platform of the *Gare du Nord*. He told me he had little confidence in the"indeterministic" interpretation, and that he was worried about the exaggerated turn that quantum physics was taking towards formalism.
>
> Then, possibly going further than he normally liked to go, he added that all physical theories, their mathematical expression apart, ought to lend themselves to so simple a description that even a child could understand it ... and what could be less simple than the purely statistical interpretation of wave mechanics. Outside the station, he left me, saying "Carry on. You are on the right track."

Louis Victor de Broglie
Courtesy of the Fondation Louis de Broglie

Despite Einstein's encouragements, Louis de Broglie abandoned his deterministic approach a few months later. We know from his reminiscences of 1966 that he left Solvay V with the discouraging feeling that he had failed to convince his colleagues – in particular Bohr, Born and Heisenberg – of the fruitfulness of the idea of an electron "guided in its movement by a pilot-wave." Louis described the situation as follows ([L. de Broglie (1966)], p. 15):

> Faced with the difficulty I encountered in giving my ideas a truly satisfactory form, and following the meeting in Brussels in October 1927 of the fifth Solvay Physics Council, where the purely probabilistic interpretation of wave mechanics, advocated by Messrs. Bohr, Born and Heisenberg, prevailed despite the objections of Einstein and Schrödinger, I was stopped in my efforts to find another.

Several factors contributed to Louis de Broglie's discouragement. On the one hand, he was influenced by Léon Rosenfeld, one of his pupils. After meeting Pauli in a corridor during Solvay V [Jacobsen (2012)], Rosenfeld moved to Göttingen, where he joined the indeterministic school. It can be argued that, after speaking with Jordan, Rosenfeld persisted in his attempt to convince his former mentor to abandon his approach.[9]

Another factor was Pauli's attitude at Solvay V. Pauli was critical of de Broglie's ideas, explaining that he did not believe that the representation introduced for material particles could be developed in a satisfactory manner. During the general discussion, Pauli expressed concern about the application of de Broglie's point of view to collision processes ([Solvay V (1928)], p. 280):

> It seems to me that Mr. de Broglie's conception with regard to the statistical results of collision experiments is in agreement with Born's theory in the case of elastic collisions, but that it is not more so when we consider non-elastic collisions.

As noticed by Bacciagaluppi and Valentini ([Bacciagaluppi and Valentini (2009)], p. 226), de Broglie's reply to Pauli's criticism contained the essential ideas needed for a proper rebuttal.

However, it emerges from later interviews that de Broglie continued to regard Pauli's criticism as a powerful argument against pilot-wave theory, even if he was "never entirely convinced of its relevance."[10] As we will see, Pauli recognised many years later that his objections were never of a fundamental character. We also know from an interview by Georges Lochak that, at the time, Louis was affected by circumstances, unrelated to science, which led him to subscribe, at least formally, to the ideas of Bohr and Heisenberg. On the one hand, his mother fell ill, and he had to take

[9] See the letters exchanged between Rosenfeld and Louis de Broglie, *Niels Bohr Archive.*
[10] See [Rocci and Lambert (2024)], p. 43

care of her. On the other, he was appointed as professor at the Sorbonne (1928), an honour that came with the obligation to teach, and prevented him from concentrating his efforts on deepening his approach.

Wolfgang Pauli (1900–1958)
Courtesy of the Nobel Foundation archive

The Pauli–de Broglie controversy resumed during the 1950's, in the light of David Bohm's deterministic theory of quantum mechanical processes.[11] However, the relations between the two men remained as cordial as ever. On February 2, 1952, Pauli wrote the following lines in a letter to de Broglie[12]: "The purpose of this letter is to discuss different scientific questions in connection with your old idea of a "pilot wave" (onde pilote), about which I had a discussion with you already at the Solvay meeting of 1927. I am always thinking back with the greatest pleasure to these old days and I beg

[11]Bohm rediscovered Louis de Broglie's approach, but the two viewpoints are different. Further information about Bohm and the relation between the two approaches can be found in [Bacciagaluppi and Valentini (2009)].

[12]Pauli's letter is part of Maurice de Broglie's *Registre contenant des pieces manuscrites concernant les premiers Congrès Solvay*, Archives de l'Académie des sciences, Paris, Institut de France.

you to show the same patience and kindness toward me which I was so glad to enjoy in the past..." In a footnote on the first page of the letter, Pauli took care to add: "I heartily agree that my old objections of 1927 at the Solvay meeting were not of a fundamental character."

II.9.4 On the Bohr–Einstein debate

A special event, usually associated with Solvay V, is the *Bohr–Einstein debate* on the status of quantum theory and the validity of Heisenberg's uncertainly relations, a debate of which one finds no trace in the Minutes of the Council.

There is no doubt that lively discussions between Einstein and Bohr took place during Solvay V at the hotel, in the morning at breakfast, or in the evening after dinner. This fact is corroborated by several witnesses, including Bohr himself, who insisted on Einstein's discomfort at the prospect of a definitive abandonment of causal descriptions of microscopic phenomena in space and time [Bohr (1959)].

Reporting many years later on Solvay V, Heisenberg provides us with a detailed description of the Bohr–Einstein controversy on quantum theory ([Heisenberg (1972)], p. 376):

> We were all staying in the same hotel,[13] and the liveliest discussions took place, not in the conference room, but during the meals at the hotel. The main protagonists of the controversies around the new interpretation of quantum theory were Bohr and Einstein. Einstein was not ready to accept the essentially statistical character of the new theory.
>
> Of course, he did not object to making probabilistic predictions when one does not know all parameters of a given system. After all, statistical mechanics and thermodynamics were based on such predictions. But what Einstein did not want to admit was the fact that it is fundamentally impossible to know all parameters necessary for a complete determination of a physical process ... He could not bring himself to accept the uncertainty relations, and tried to imagine experiences where these relations could not be applied.

[13]Contrary to a wide-spread belief, the members were not hosted at Hotel Métropole, but at *Grand Hôtel Britannique*.

> Our controversies usually began early in the morning, Einstein coming up at breakfast with an ideal experiment, likely, in his opinion, to contradict the uncertainty relations. We immediately started analysing the experiment, and, on our way to the conference room – I used to accompany Bohr and Einstein – a first clarification of the question asked and the statement made was carried out.
>
> During the day, many discussions were held on this problem, and in general we arrived in the evening at a point where Bohr could prove to Einstein, over dinner, that the proposed experiment could not lead to a refutation of the uncertainty relations. Einstein was then slightly worried, but already the next morning, at breakfast, he had another ideal experiment ready to offer, more complicated than the previous one, and, in his opinion, likely to definitively undermine the uncertainty relations. This attempt was to fail the same day, just like the previous one.

In 1961, at the twelfth Physics Council, Bohr recalled the debates he had with Einstein: at Solvay V (1927), in the presence of Ehrenfest and under his influence ([Solvay XII (1962)], p. 26) and at Solvay VI (1930), when Einstein brought up new arguments to resume the discussion of the epistemological problems debated at the fifth Council.

On another occasion, Bohr described the focus of their discussions at Solvay V ([Bohr (1961)], p. 160):

> The central point was to decide whether the quantum-mechanical description exhausted all possibilities of accounting for observable phenomena, or whether, as Einstein maintained, the analysis could be carried further, and a fuller description of the phenomena could be obtained by bringing into consideration the detailed balance of energy and momentum in individual processes.

As indicated, the Bohr–Einstein controversy did not end at Solvay V. It took a more dramatic turn at Solvay VI. During this last Council attended by Einstein, Bohr managed, after a restless night, to detect a flaw in a well-known scenario, imagined by Einstein in an attempt to defeat the uncertainty principle.[14]

[14]Einstein made use of a clock and relied on the equivalence between mass and energy, a well-known consequence of the special theory of relativity. However, Bohr managed to refute Einstein's argument by noting that the effect of a gravitational field on the clock, a consequence of Einstein's general relativity theory, had not been taken into account.

Einstein and Bohr discussing during Solvay VI in 1930
Courtesy of the Niels Bohr archive, Copenhagen

However, when reporting on the Solvay meetings three decades later, Bohr recognised the importance of the influence that the discussions with Einstein had exerted on his thinking, allowing him to grasp the true meaning and scope of the complementarity principle.

At Solvay V, despite their awareness that several questions were still open to debate, the members of the *indeterministic school* were confident that they had succeeded in convincing almost all their colleagues of the need to embrace their point of view. Here is Heisenberg's description of the role of Solvay V in the general acceptance of the "Copenhagen interpretation" of quantum theory ([Heisenberg (1972)], p. 378):

In the autumn of 1927 the Solvay Conference took place in Brussels, and here the new interpretation of quantum theory was exposed to the most ingenious criticism, particularly on the part of Einstein, and thereby received the crucial test. Again experiments were discussed whose interpretation had always offered the greatest difficulties, and it became apparent over an over again that the new interpretation contained no internal contradictions, and clearly led to the correct experimental results. One of the first documents of the discussion at this conference is Bohr's article for Albert Einstein's 70th birthday. Since the Solvay Conference of 1927 the Copenhagen interpretation has been fairly generally accepted and has formed the basis of all practical applications of quantum theory. It has, however, occasionally been contradicted and criticised as the orthodox theory.

Epilogue

H. A. Lorentz during Solvay V
in front of the Solvay Institute of Physiology
Courtesy of the Niels Bohr archive, Copenhagen

Lorentz died on February 4, 1928, less than ten weeks after the end of Solvay V. His outstanding presidency of the *International Scientific Committee* (ISC) remained alive in the minds of all those who took part in the *historic meetings* that we discussed in this book. As indicated in the Introduction, our purpose was not to present a comprehensive account of the steps that led to the formulation of quantum mechanics. Our aim was to show that the process which marked the transition from classical to quantum physics took place during Lorentz's presidency of five Physics Councils, and was, to a great extent, influenced by Solvay's convening of the 1911 Council, and his subsequent founding of an *International Physics Institute* (ISIP).

We also wanted to emphasise a point that for a long time, escaped the attention of science historians: the fact that ISIP's subsidy programme, set up at Solvay's request, had a direct impact on major advances in the development of quantum theory. Some of them were realised before WWI by physicists who took part in the first two Physics Councils, or benefited from Solvay grants. Later developments, were influenced by events and discussions that took place during the post-war Councils chaired by Lorentz.

Looking back at the stages described in Parts I and II, it appears that the *first* quantum revolution was the result of a process that started at Solvay I, in the light of a general awareness that Planck's *quantum of action* had to be taken seriously, and ended at Solvay V with the birth of quantum mechanics. As for the intermediate steps, discussed at Solvay II, III and IV, in public or in private, they were taken after realising that the quantum of action, introduced in the context of radiation, should also play a role in the structure of atoms.

Two crucial advances, discussed at Solvay II, or on its sidelines, concerned the undulatory nature of X-rays, exhibited by Laue's diffraction spots, and Bohr's atomic model, governed by the quantum of action.

Solvay III was the Council at which Maurice de Broglie's X-ray results established the validity of the Bohr-Rutherford atomic model. Thanks to its influence on Compton's research, Maurice de Broglie's work of 1922 opened the way to a general recognition of the dual nature of radiation. It also encouraged his brother Louis to delve into research that allowed him to extend the "wave-particle" duality of light, and predict the existence of matter waves.

Solvay IV, although not focusing on quanta, provided elements in favour of Einstein's light-quantum interpretation of the photoelectric effect, notably thanks to the result of the Joffé–Dobronravoff experiment. It also enabled Langevin to discuss Louis de Broglie's doctoral thesis "in private" (a discussion that may have been attended by Schrödinger, as suggested by Joffé).

As for Solvay V, it was highlighted by Compton's *photon theory*, and by three proposals for a new mechanics: Louis de Broglie's *pilot wave* theory, the Born–Heisenberg *matrix mechanics*, and Schrödinger's *wave mechanics*.

Thanks to the presence of Charles Thomson Rees Wilson, the Council paid tribute to the man whose expansion method played a decisive role in the quantum revolution. Indeed, as indicated in this book, major steps in the development of quantum physics would not have taken place if physicists had not been able to rely on Wilson's photographs.

C. T. R. Wilson at Solvay V
sitting in front of Max Born and Niels Bohr
Courtesy of the Solvay Institutes

We therefore decided to devote two appendices to the invention of the cloud chamber, and to the recognition of its importance by the Nobel Committee for physics. With regard to this last point, it is worth emphasising that it also took sixteen years for the Nobel Foundation to recognise the merits of Wilson's cloud chamber, a device, developed in 1911, which for the first time exhibited the spectacular effects of the passage of an ionising particle through a gas. It was this device that allowed Rutherford to confirm his theories. The power of the Wilson method appeared most clearly in the context of Compton's *photon theory*. By confirming the existence of recoil electrons, it established beyond any doubt the reality of the Compton effect. Last but not least, Wilson's photographs provided Heisenberg with evidence that allowed him to formulate the *indeterminacy relations* which characterise the quantum world.

A few weeks after the fifth Solvay Council, Compton and Wilson shared the 1927 Physics Nobel Prize: Compton for discovering the "effect"; Wilson for "his method of making the paths of electrically charged particles visible by condensation of vapour."[15] The complementary merits of Compton and Wilson were highlighted by Manne Siegbahn in his "Presentation Speech" at the Nobel Award ceremony.[16] The shared Compton–Wilson Prize was an official recognition of the wave-particle nature of light, a fact suggested twenty-two years earlier by Einstein, extended to matter in by Louis de Broglie in 1924, and highlighted at Solvay V with Bohr's complementarity principle.

As a conclusion, we should say that we share Bohr's feeling, as he ended his report to the twelfth Physics Council in 1961 with the following words ([Solvay XII (1962)], p. 36):

> I want to express the hope that this review of some features of the historical development may have given an indication of the debt which the community of physicists owe to the Solvay Institute, and of the expectations we all share for its future activities.

The former Solvay Institute of Physiology (Parc Léopold – Brussels), main stage of four of the Councils chaired by Lorentz.

Courtesy of the Solvay Institutes

[15]The website of the Nobel Foundation also contains a summary of the path that led to Wilson's invention of the cloud chamber.

[16]See Appendix D.

Acknowledgements

We wish to thank the *International Solvay Institutes for Physics and Chemistry* founded by Ernest Solvay, and in particular the President *Mr. Jean-Marie Solvay* and the Director *Professor Marc Henneaux* for their encouragements and support in the writing of this book.

Professor Marc Henneaux also deserves our special thanks for his valuable advice and assistance in getting the book published by World Scientific, the official publisher of the current Proceedings of the Physics and Chemistry Councils.

We would like to express our gratitude to *Professor Jan Danckaert*, Rector of the Vrije Universiteit Brussel (VUB) for his interest in our project, and for the warm welcome he gave to Alessio Rocci, by securing him a postdoctoral fellowship within the framework of the special relationship between the Brussels Universities and the International Solvay Institutes. We also wish to acknowledge the support extended to Dr. Rocci at the Université Libre de Bruxelles (ULB), by *Professor Kenneth Bertrams*, expert in the History of Science & Technology and connoisseur of the history of the Solvay Company, and by *Director Renaud Bardez*, responsible for the ULB Archives, Heritage and Precious Collections.

This work would not have been possible without the generous support of *Dr. Krist'l Van Ouytsel*, which enabled the Institutes to recruit a historian of science, responsible for the reorganisation and enhancement of their archives within the context of the *Solvay Science Project.*[1] It was Krist'l's dedication to this project, and the means that she made available for its realisation, that resulted in the appointment of Dr. Rocci as

[1] `http://www.thesolvayscienceproject.be/exhibits/show/the-solvay-science-project?lang=en`

postdoctoral fellow at VUB through the *Fonds voor natuurwetenschappen in de samenleving* (Fund for natural sciences in society).[2] .

It must be added that it was also thanks to the moral and financial support of the Solvay family – more specifically of Mr. Jean-Marie Solvay and Mrs. Marina Solvay – that Dr. Rocci's appointment could be extended so as to allow the continuation of the collaboration that gave rise to this book. We hope that these enlightened patrons will find here a testimony of our gratitude.

We also express special thanks to *Mrs. Dominique Bogaerts* and *Mrs. Isabelle Van Geet*, from the Institutes, for the never ending moral and practical support they extended to us during all these years.

With regard to our research, we would like to thank *Professor Frits Berends* for communicating important archival documents regarding Ehrenfest, Lorentz's successor at Leiden University, and *Professor Michael Eckert*, for providing us with decisive information on Erwin Schrödinger's work of 1924 on pre-quantum mechanical electron gas theories.

We also thank Mariska Krikken from the *Noord-Hollands Archief* (North-Holland Archives) in Haarlem (NL), for providing digital copies of H. A. Lorentz's correspondence.

We acknowledge Christian Joas, the director of the Niels Bohr Archive in Copenhagen (NBA), and the archivist Rob Sunderland for providing us the pictures owned by NBA published in this book. For the same reason, we also thank Ulrika Magnusson of the Nobel Foundation.

One of us, Franklin Lambert, is particularly grateful to Mr. Clovis Neves for helping him to overcome some of the technical difficulties generated by an increasingly digital world.

As for Alessio Rocci, he is grateful to *Professor Emeritus Franklin Lambert*, co-author of this book, for his endless patience in our discussions about the history of the International Solvay Institutes and his endless support in exploring the history of physics. A.R. also thanks: *Professor David Gross*, chair of the present International Scientific Committee for the Physics Councils, for his stimulating questions about Lord Rutherford's attitude; *Professor Kenneth Bertrams* and *Professor Brigitte Van Tiggelen* for their suggestions during the postdoctoral years in Brussels; the team of the ULB archive *Réserve précieuse* for their help. A.R. is grateful to

[2]`https://www.vub.be/en/support-vub/support-vub-through-gift/fund-for-natural-sciences-in-society`

Professors Giulio Peruzzi and Kurt Lechner for their advice and friendship. This book is also dedicated to the loving memory of Professor Mario Tonin, who was the first to make A.R. understand and appreciate the beauty of Quantum Mechanics. A.R. is deeply indebted to his wife, *Sara Elisabetta Maria Minuto*, for her continuing encouragement over the postdoctoral years he had to live far way from her.

Appendix A

Joffé's memories of Solvay IV

The following quotation (translated in English) has been taken from A. F. Joffé's book *Encounters with Physicists*[1] [Joffé (1967)]:

> The Belgian chemist Solvay, who invented a process for producing soda, built several factories that brought him large amounts of income. He decided to use part of it to solve some fundamental questions in physics and chemistry that interested him. For this purpose, an international committee was created, whose chairman was the outstanding representative of theoretical physics, Hendrik Lorentz. The Solvay Congress met for the first time in 1911. All the leading scholars of the time took part. After the World War from 1914 to 1918, the Congresses met every three years. A year and a half before the convening of each Congress, the International Solvay Committee selected a narrow problem in physics and eight scholars who could possibly solve it best. Also invited were 10 to 15 physicists whose participation could be beneficial.
>
> For eight days, different sides of the same problem were discussed at the sessions of the Congress, which lasted from 10 a.m. to 6 p.m., with a short break for a light lunch. Initially, Lorentz led the congresses, then Langevin, and after the Second World War, Bragg. Each Congress brought decisive progress to the problem that was addressed ... At the time (1924), the Solvay Congresses took place under the high patronage of King Albert, who had made a name for himself by resisting the invading German troops in the

[1] The German translation of the original version published in Moscow in 1962 reads: *Begegnungen mit Physikern*.

First World War. The King received the members of the Congress at his court. At the reception, I stood in a row with Debye and Schrödinger. When the King approached us, he first talked to Schrödinger, who was then a professor in Switzerland. The conversation covered the educational system in this country. The King asked me about Leningrad University and how atomic theories explained Mendeleev's system. Finally, he turned to Debye. After the collapse of Germany and Austria at the end of Word War I, both Debye and Schrödinger moved to Switzerland – to Zurich, where living conditions were better. The entire store of knowledge that the King had at his disposal had obviously been exhausted in the conversation with Schrödinger. The King tried to move the conversation to another country by asking where Debye had studied. Debye came from Aachen with Sommerfeld. But at that time, immediately after the war, it did not seem appropriate to mention Germany. Debye replied that he first studied technology and then physics. The King could not find a way out for a long time, but then Lorentz came to help. He assessed the situation with his characteristic insight, walked up, and said "There's Professor Debye, our Dutch colleague", and everything went normally. The incident shows Lorentz's quick-wittedness and his ability to find ways out of difficult situations. I turned to him several times when I encountered obvious contradictions or when I wanted to check my physical ideas.

By the way, I have to say that King Albert proved to be a cultured and educated person. Having been elected member of the Solvay Committee, I met him several times. I remember the King's lively interest in the influence of the unification of large collective state farms in the Soviet Union on the agricultural technology, his strong opposition to war, and his humorous remarks about the situation of a King in a constitutional country invited to talk about the prime minister's opinion.

Appendix B

Pais about Bohr, Compton and the BKS proposal

The following lines are excerpts from chapter 11 of Abraham Pais's book [Pais (1991)]:

> The first time that Bohr and Einstein found themselves in scientific opposition goes back to 1924, the final year of the old quantum theory ... At issue was their difference of opinion concerning the reality of light quanta ... Several questions arise. First, why was Bohr opposed to the photon? Note to begin with that in his earlier work on atomic structure Bohr had discussed emission and absorption of light in atomic transitions without, however, examining the mechanisms for these processes, nor the nature of light itself. Note further that Bohr was but one among numerous leading theoreticians (not to mention experimenters, such as Millikan) to take a stand against the photon. Their principal common reason was that Maxwell's theory of wave propagation of electromagnetic fields had been thoroughly confirmed by experiment and was considered both as one of the most simple and one of the most solid parts of physical theory. Quantum phenomena had cast doubts on classical theory but, it was widely believed, these had only to do with interactions between matter and radiation – an obscure region of theory in any event – not with radiation itself ...
>
> Secondly, what was the status of the energy conservation principle in 1923, when Bohr put it in doubt? In the domain of macroscopic physics this principle had occupied the status of a fundamental law ever since the mid-nineteenth century. For what follows it is essential to realise, however, that in 1923 this law

had never yet been experimentally tested at the level of individual microscopic processes such as atomic transitions, collisions of electrons with electrons or atoms, etc. ... The first verifications that energy and momentum conservation do hold at the microscopic level date from 1925. To this day these laws are believed to hold strictly.

Thirdly, what good could it possibly do for the theory of light and matter to abandon energy conservation? The answer is that energy non-conservation is inevitable if one were to accept the odd, in fact false, dichotomy held by the anti-photon camp that in emission or absorption of light the energy of material, atomic, systems changes discontinuously, in the quantum way, while the (electromagnetic) energy of light changes continuously in the classical way.

Bohr was not the first to contemplate energy non-conservation as a means for avoiding light quanta. Already in 1910, Einstein had briefly toyed with that idea but had rejected it. In 1916, Nernst published an article "On an attempt to revert from quantum mechanical considerations to the assumption of continuous energy changes", in which he proposed that energy was only conserved statistically, that is, when averaged over an assembly of individual processes of a specific kind. This paper influenced Bohr, as is seen from his unpublished manuscript written in 1917 or 1918: "It would seem that any theory capable of an explanation of the photoelectric effect as well as interference phenomena must involve a departure from the ordinary theorem of conservation of energy as regards the interaction between radiation and matter. This view has been expressed by several authors and an interesting attempt to build a theory on this basis has been made by Nernst". In 1919, Darwin, a friend of the Manchester days, wrote to Bohr "At present I consider the case against conservation quite overwhelming" ... Also Sommerfeld remarked that the"mildest cure" for reconciling the wave with the particle picture would be to abandon energy conservation. The situation changed drastically in 1923, when a crucial high-quality experiment by Arthur Compton created a sensation among physicists of that time . He had observed the "Compton effect": when light is scattered by electrons, its frequency diminishes by an amount that depends on the angle of scattering. Compton found this change to be in agreement with the predictions (made by himself and independently by Debye) that follow from the assumptions, first that a light beam behaves as a bundle of light quanta, secondly, that energy and momentum are conserved in light quantum- electron scattering. In 1924, Sommerfeld called this result "probably the most important discovery that could have been made in the current state of physics."

It seemed that the issue of possible non-conservation of energy had been laid to rest. Bohr, Kramers, and Slater (BKS) thought otherwise however ... Regarding Compton's results, BKS noted, correctly, that so far his experiments had only confirmed energy-momentum conservation averaged over many individual processes, which was not in conflict with their statistical viewpoint ... In the early months of 1925, two experiments showed conclusively that BKS had been on the wrong track. Walther Bothe and Hans Geiger developed counter coincidence techniques for the purpose of checking whether in the Compton effect the secondary photon and electron are produced simultaneously, as causality demands. They found that these two particles are both created within a time interval less than 10-8 seconds. Causality had been established and the randomness of the relative moments of creation demanded by BKS disproved.

Appendix C

The Wilson "cloud chamber"

The following lines are part of a Biographical Note on C.T.R. Wilson, reproduced from the website of the Nobel Foundation.

> When standing on the summit of Ben Nevis, the highest of the Scottish mountains, in the late summer of 1894, Wilson was struck by the beauty of coronas and "glories" (coloured rings surrounding shadows cast on mist and cloud), and he decided to imitate these natural phenomena in the laboratory (early 1895). His sharp observation and keen intellect, however, led him to suspect (after a few months' work at the Cavendish Laboratory) that the few drops reappearing again and again each time he expanded a volume of moist, dust-free air, might be the result of condensation on nuclei — possibly the ions causing the "residual" conductivity of the atmosphere – produced continuously. Wilson's hypothesis was supported after exposure (early 1896) of his primitive cloud chamber to the newly discovered (end of 1895) X-rays. The immense increase of the "rain-like" condensation fitted excellently with the observation made by Thomson and McClelland immediately after Röntgen's discovery, that air was made conductive by the passage of X-rays. When, during the summer of that year, it was firmly established by Thomson and Rutherford that the conductivity was indeed due to ionisation of the gas, there was no longer any doubt that ions in gases could be detected, photographically recorded and thus studied at leisure (...). Early in 1911, he (Wilson) was the first person to see and photograph the tracks of individual α- and β-particles and electrons

... The event aroused great interest as the paths of the α-particles were just as W. H. Bragg had drawn them in a publication some years earlier. But it was not until 1923 that the cloud chamber was brought to perfection and led to his two, beautifully illustrated, classic papers on the tracks of electrons.

Appendix D

Excerpt from Siegbahn's *Nobel Presentation Speech* (December 10, 1927)

The Royal Academy of Sciences has awarded this year's Nobel Prize in Physics to Professor Arthur Holly Compton from the University of Chicago for the discovery of the phenomenon named after him the Compton effect, and to Professor Charles Thomson Rees Wilson of the University of Cambridge for his discovery of the expansion method of rendering visible the tracks of electrically charged particles.

Professor Compton won his prize by work in the field of X-radiation. Soon after Röntgen's discovery it became clear that matter exposed to X-rays emits radiations of different character. Besides an emission of electrons, corresponding to the photoelectric effect, known also in the optical region of radiation, there is also a secondary X-radiation. Even before the methods of X-ray spectrometry were known, these secondary X-rays were proved, by the investigation of their absorption, to be of a twofold nature. It was Barkla who, through his fundamental researches, proved that the secondary X-radiation consists partly in scattering of X-rays, which he thought to have the same penetrability as the original radiation, and partly in a specific radiation which was characteristic of the chemical atom, and which was more easily absorbed. When X-rays fell upon matter, with small atomic weight, as for example graphite, Barkla was not able to detect the mentioned characteristic X-radiation, but only a scattering; and consequently the secondary X-rays ought to have the same properties as the original X-rays. Barkla, however, had already been able to show that in this case also the secondary X-rays – at least partially – are more easily absorbed than the original radiation, and therefore have a greater wavelength. Barkla thought this to be a new characteristic X-radiation. This is the point where Compton comes in and affects the development of science. He made exact spectrometric investigations of the secondary X-radiation from matter with small atomic weight: in other words, he undertook to investigate exactly the scattered radiation. After some preliminary work, he found an experimental method that gave results that were as exact as astonishing. Using homogeneous X-rays – corresponding optically to monochromatic illumination, that is to say, to the use of a source of light that emits only one single spectral line – he found that the scattered radiation consists of two lines, one exactly the same as that of the source of rays, the other with a somewhat greater wavelength. This is the first evident manifestation of the Compton effect. Its reality was at first disputed, but of later years it has been well established and verified.

The change in wavelength soon proved to be independent of the nature of the matter used for scattering, while it varies with the angle between the incident and the scattered rays. Hence, the phenomenon cannot be explained as a new characteristic radiation of the same nature as that hitherto known; and Compton deduced a new kind of corpuscular theory, with which all experimental results showed perfect agreement within the limits of experimental error. According to this theory, a quantum of radiation is re-emitted in a definite direction by a single electron, which in so doing must recoil in a direction forming an acute angle with that of the incident radiation. In its mathematical dress, this theory leads to an augmentation of the wavelength that is independent of the wavelength of the incident radiation, and implies a velocity of the recoil electron that varies between zero and 80% of the velocity of light, when the angle between the incident and the scattered radiation varies between 0° and 180°. Thus, this theory predicts recoil electrons with a velocity generally much smaller than that of the above-mentioned electrons which correspond to the photoelectric effect. It was a triumph for both parties when these recoil electrons were discovered by Wilson himself and, independently, by another investigator. Hereby the second chief phenomenon of the Compton effect was experimentally verified, and all observations proved to agree with what had been predicted by Compton's theory.

Bibliography

Andrade, E. N. C. (1964). *Rutherford and the nature of the atom* (Doubleday & Company Inc., New York).

Arabatzis, T. (2004). The Zeeman Effect and the Discovery of the Electron. In: Buchwald, J.Z. and Warwick, A. (2004). *Histories of the electron: the birth of microphysics* (MIT Press, Cambridge – MA).

Authier, A. (2013). *Early Days of X-Ray Crystallography* (Oxford Academic Press, Oxford – UK).

Bacciagaluppi, G. and Valentini, A. (2009). *Quantum Theory at the Crossroads:Reconsidering the 1927 Solvay Conference* (Cambridge University Press, Cambridge – UK).

Badash, L. (1979). The Origin of Big Science: Rutherford at McGill. In: Bunge, M. and Shea, W.R. (Eds). *Rutherford and Physics at the Turn of the Century* (Dawson, Folkestone, Kent, England, and Science History Publications).

Baracca, A. (2005). Mechanistic Science, Thermodynamics and Industry at the End of the Nineteenth Century. In: *Physics Before and After Einstein*, M. Marmone Capria (Ed.) (IOP Press, Washington, DC).

Barkan, D.K. (1993). The Witches' Sabbath. The First International Solvay Congresses in Physics. In: Beller, M., Cohen, E.S., and Renn J. (Eds.) *Einstein in Context* (Cambridge University Press, Cambridge – UK).

Barkla, C. G. and Collier, G. (1912). The Absorption of X-rays and Fluorescent X-Ray Spectra, *Philosophical Magazine*, **23**, pp. 987–997.

Barkla, C. G. and Martyn, G. (1913a). Interference of Röntgen radiation, *Proceedings of the Physical Society of London*, **25**, pp. 206–213.

Barkla, C. G. and Martyn, G. H. (1913b) The Photographic Effect of X-Rays and X-Ray Spectra, *Philosophical Magazine*, **25**, pp. 296–300.

Barkla C. G., and Philpot A.J. (1913c) Ionisation in gases and gaseous mixtures by Röntgen and corpuscular (electronic) radiations, *Philosophical Magazine*, **25**, pp. 832–856.

BAAS meeting (1913). Physics at the British Association, *Nature* **92**, pp. 304–309.

Bell J. S. and Aspect A. (2004). *Speakable and Unspeakable in Quantum Mechanics: Collected Papers on Quantum Philosophy*, 2nd ed. (Cambridge University Press, Cambridge – UK).

Berends, F. (2014). *Lorentz' gouden doctoraat: een huldiging zonder weerga.* Series: *Verhalen rond een foto* (Koninklijke Hollandsche Maatschappij der Wetenschappen, Haarlem).

Bohr, N. (1913a). On the theory of the decrease of velocity of moving electrified particles on passing through matter, *Philosophical Magazine*, **25**, pp. 10–31.

Bohr, N. (1913b-c-d) On the Constitution of Atoms and Molecules, Pat.I-II-III *Philosophical Magazine*, **26**, pp. 1–24; pp. 476–502; pp. 857–875.

Bohr, N. (1959). Discussion with Einstein of Epistemological problems in atomic physics. In: Schlipp, P.A. (1959). *Albert Einstein: Philosopher-Scientist. The Library of Living Philosophers. Vol. VII* (Harper & Row, New York).

Bohr, N. (1961). The Rutherford Memorial Lecture 1958. Reminiscences of the Founder of Nuclear Science and of Some Developments Based on his Work, *Proceedings of the Physical Society* **78**, p. 1083.

Bohr, N., Kramers H. A., and Slater, J.C. (1924). The Quantum Theory of Radiation, *Philosophical Magazine*, **47**, pp. 69–87.

Bolles, E. B. (2004). *Einstein Defiant. Genius versus Genius in the Quantum Revolution.* (Joseph Henry Press, Washington, DC).

Born, M., Heisenberg, W. and Jordan, P. (1926). Zur Quantenmechanik II, *Zeitschrift für Physik* **35**, pp. 557–615.

Born, M. and von Karman, T. (1912). Zur Theorie der Spezifischen Wärmen, *Physikalische Zeitschrift* **14**, pp. 15–19.

Born, M. and Jordan, P. (1925). Zur Quantenmechanik. *Zeitschrift für Physik*, **34**, pp. 858–888.

Born, M. (2004). *The Born-Einstein Letters, 1916-1955. Friendship, Politics and Physics in Uncertain Times* (Macmillan Science, London).

Bose, S. N. (1924). Planck's Gesetz und Lichtquantenhypothese, *Zeitschrift für Physik*, **26**, pp. 178–181.

Bothe, W. and Geiger, H. (1924). Ein Weg zur experimentellen Nachprüfung der Theorie von Bohr, Kramers und Slater, *Zeitschrift für Physik* **26**, p. 44.

Boya, L. J. (2003). The Thermal Radiation formula of Planck (1900), *Revista de la Real Acedemia de Ciencias, Zaragoza* **58**, pp. 91–114.

Casimir, H. (1955). On the theory of superconductivity. In: Pauli, W. (Eds) *Niels Bohr and the Development of Physics: Essays Dedicated to Niels Bohr on the Occasion of his Seventieth Birthday* (Pergamon Press, Oxford – UK).

Cockcroft, J. D. (1967). George de Hevesy 1885-1966, *Biographical Memoirs of Fellows of the Royal Society*, **13**, pp. 125–126.

Coffey, P. (2008). *Cathedrals of Science. The Personalities and Rivalries that Made Modern Chemistry* (Oxford University Press, Oxford – UK).

Compton, A. H. (1922). Secondary radiations produced by X-rays and some of their applications to physical problems, *Bullettin of the National Research Council*, **20**, pp. 1–56.

Compton, A. H. (1961). The Scattering of X rays as particles, *American Journal of Physics*, **29**, pp. 817–820

Crawford, E. (1987). *The Beginnings of the Nobel Institution. The Science Prizes, 1901-1915* (Cambridge University Press, Cambridge – UK).

Darrigol, O. (1993). Strangeness and Soundness in Louis de Broglie's Early Work, *Physis*, **30**, pp. 303–372.

Darwin, C. G. (1955). The Discovery of Atomic Number. In: Pauli, W. (Ed.) *Niels Bohr and the Development of Physics: essays dedicated to Niels Bohr on the occasion of his seventieth birthday* (Pergamon Press, Cambridge – UK).

De Broglie, L. (1920). Sur l'absorption des rayons X par la matière, *Comptes-rendus de l'Académie des Sciences*, **171**, pp. 1137–1139.

De Broglie, L. (1921). Sur la théorie de l'absorption des rayons X par la matière et le principe de correspondence, *Comptes-rendus de l'Académie des Sciences*, **173**, pp. 1456–1458.

De Broglie, L. (1922). Rayons X et équilibre thermodynamique, *Journal de Physique*, **3**, pp. 33–45.

De Broglie, L. (1931). *Notice sur les Travaux Scientifiques de M. Louis de Broglie* (Hermann & Co., Paris).

De Broglie, L. (1947). *Physique et Micro physique*, (Albin Michel, Paris).

De Broglie, L. (1953). Vue d'ensemble sur mes travaux scientifiques. In: De Broglie, L. *Physicien et penseur*, (Albin Michel, Paris).

De Broglie, L. (1966). *Certitudes et incertitudes de la science* (Albin Michel, Paris).

De Broglie, L. (1976). My meeting with Einstein at the Solvay Conference of 1927. In: French, A.P. (Ed.). (1976). *Einstein: A Centenary Volume* (Harvard University Press, Cambridge – MA).

De Broglie, M. (1908). *Recherches sur les centres électriques de faible mobilité dans les gaz* (Gauthier-Villars, Paris).

De Broglie, M. (1911). *Registre contenant des pièces manuscrites concernant les premiers Congrès Solvay, 2 volumes*, Archives de l'Académie des sciences, Paris, Institut de France (unpublished).

De Broglie, M. (1921). Les phénomènes photo-électriques pour les rayons X et les spectres corpusculaires des éléments, *Journal de Physique et le Radium* **2**(9), pp. 265–287.

De Broglie, M. (1921b). Sur les spectres corpusculaires et leur utilisation pour l'étude des spectres des rayons X, *Comptes-rendus de l'Académie des Sciences*, **173**, 1156–1161.

De Broglie, M. (1922). *Les rayons X*, (Librairie Scientifique Albert Blanchard).

De Broglie, M. (1951). *Les Premiers Congrès Solvay et l'orientation de la Physique depuis 1911*, (Albin Michel, Paris).

P. Debye, P. (1912). Zur Theorie der spezifischen Wärmen, *Annalen der Physik*, **344**, pp. 789–839.

Dirac, P. A. M. (1925). The fundamental equations of quantum mechanics, *Proceedings of the Royal Society of London A* **226**, pp. 642–653.

Dirac, P. A. M. (1978). *Lectures delivered at the University of New South Wales in 1975* (Directions in Physics – Wiley, New York).

Dunoyer,L. (1913). Sur l'aberration de sphéricité dans les objectifs, *Journal de Physique Theorique et Applquée*, **3**(1), pp. 468–486.

Eckert, M. (1989). Das "frei Elektronengas". Vorquantenmechanische Theorien über die elekltrischen Eigenschaften der Metalle. In: *Wissenschaftliche Jahrbuch des Deutschen Museums*, pp. 57–91.

Eckert, M. (2012). Disputed discovery: the beginnings of X–ray diffraction in crystals in 1912 and its repercussions, *Acta Crystallographica* **A68**, pp. 30–39.

Eckert, M. (2012). Max von Laue and the Discovery of X-Ray Diffraction in 1912, *Annalen der Physik* **524**, pp. A83–A85.

Eckert, M. (2014). How Sommerfeld extended Bohr's model of the atom (1913-1916), *The European Physical Journal H*, **39**, pp. 141–156.

Eckert, M. (2015). From X-rays to the h-hypothesis: Sommerfeld and the early quantum theory 1909-1913, *The European Physical Journal H* **224**, no. 10, pp. 2057–2063.

Eckert, M. (2020) *Establishing Quantum Physics in Munich. Emergence of Arnold Sommerfeld's Quantum School*, Series: SpringerBriefs in History of Science and Technology, (Springer Cham, Cham).

Eckert, M. and Märker, K. (2004). *Arnold Sommerfeld Wissenschaftlicher Briefwechsel. Band 2*, (Diepholz-München, Munich).

Einstein, A. (1905). On a heuristic point of view concerning the production and transformation of light. In: *Collected Papers of Albert Einstein, Vol. 2*, p. 99. Translation of *Annalen der Physik* **17**, 1905, pp. 135–148.

Einstein, A. (1909). Zum Gegenwärtigen Stande des Strahlungsproblems, *Physikailische Zeitschrift* **10**, pp. 185–193.

Einstein, A. (1925). Quantentheorie des einatomigen idealen Gases. Tweite Abhandlung, *Sitzungsberichte der Preussischen Akademie der Wissenschaften (Berlin), Physikalisch-mathematische Klasse*, **1925**, pp. 3–14.

Einstein, A. and Stern, O. (1912). Einige Argumente für die Annahme einer molekularen Agitation beim absoluten Nullpunkt, *Annalen der Physik* **345**, pp. 551–560.

Einstein, A. and Podolsky, B. and Rosen, N. (1935) Can Quantum-Mechanical Description of Physical Reality Be Considered Complete? *Physical Review*, **47**, pp. 777–780.

Eve, A.S. (1939). *Rutherford*, (Cambridge University Press, Cambridge – UK).

Ewald, P.P. (1962). *Fifty years of X-ray diffraction* (NVA Oosthoek's Uitgeversmaatschappij, Utrecht).

Feather, N. (1977). Some episodes of the alpha-particle story, 1903-1977, *Proceedings of the Physical Society of London*, **357**, pp. 117–129.

Friedrich, B. (2019). Book Review: For Science, King, and Country: The prodigious life and untimely death of "Harry" Moseley, *Angewandte Chemie International Edition* **58**, pp. 2–4.

Franck, J. and Hertz, G. (1914). Über ZusammenstÖße zwischen Elektronen und den Molekülen des Quecksilberdampfes und die Ionisierungsspannung desselben, *Verhandlungen der Deutschen Physikalischen Gesellschaft* **16**, 457–467.

Freire, O. J. (2015). *The Quantum Dissidents. Rebuilding the Foundations of Quantum Mechanics (1950-1990)* (Spriger-Verlag, Berlin).

Jammer, M. (1989). *The Conceptual Development of Quantum Mechanics.* The History of Modern Physics, 1800-1950 (Series), **Vol. 12**, (Tomash Publishers and The American Institute of Physics, College Park – MD).

Jensen, C. (2000). *Controversy and Consensus: Nuclear Beta Decay 1911-1934* (Springer Basel AG – Birkäuser, Basel).

Jacobsen, A. S. (2012). *Léon Rosenfeld: Physics, Philosophy, and Politics in the Twentieth Century* (World Scientific, Singapore).

Joffé, A. F. (1947). A la Mémoire d'un Maître et Ami, *La Pensée, Revue du Rationalisme Moderne*, **12**, pp. 15–17.

Joffé, A. F. (1967). *Begegnungen mit Physikern* (B. G. Teubner Verlagsgesellschaft, Leipzig) – *The original version was published in Moscow in 1962.*

Joffé, A. F. and Dobronravoff, N. (1925). Beobachtungen über die Ausbreitung von Röntgenimpulsen, *Zeitschrift für Physik* **34**, pp. 889–892.

Heilbron, J. L. (1966). The Work of H. G. J. Moseley, *Isis* **57**, pp. 336–364.

Heilbron, J. L. (1974). *H.G.J. Moseley: The Life and Letters of an English Physicist, 1887-1915* (University of California Press, Berkeley).

Heilbron, J. L. (1977). Lectures on the history of atomic physics 1900-1922. In: Weiner, C. (Eds.) *History of Twentieth Century Physics. Proceedings of the International School of Physics 'Enrico Fermi', Course 57* (Academic Press, Cambridge – MA).

Heilbron, J. L. (1985). J.J. Thomson and the Bohr Atom. In: Weart, S.R. and Phillips, M. (Eds.)*History of Physics. Readings from Physics Today* (The American Institute of Physics, College Park MD).

Heilbron, J. L. (2015). British participation in the first Solvay Councils on Physics, *The European Physical Journal H* **224**, no 10, pp. 2041–2056.

Heilbron, J. L. and Kuhn, T. S. (1969). The Genesis of the Bohr Atom, *Historical Studies in the Physical Sciences*, **1**, pp. 211–290.

Heilbron, J. L. and Rovelli, C. (2023). Matrix mechanics mis-prised: Max Born's belated nobelisation, *The European Physical Journal H* **48**, article number 11.

Heisenberg, W. (1925). Über quantentheoretische Umdeutung kinematischer un mechanischer Beziehungen, *Zeitschrift für Physik* **33**, pp. 879–893.

Heisenberg, W. (1929). Die Entwicklung der Quantentheorie. 1918-1928, *Die Naturwissenschaften*, **17**, pp. 490–496.

Heisenberg, W. (1955). The development of the interpretation of the quantum theory. In: Pauli, W. (Eds) *Niels Bohr and the Development of Physics: Essays Dedicated to Niels Bohr on the Occasion of his Seventieth Birthday* (Pergamon Press, Cambridge – UK).

Heisenberg, W. (1969). *Der Teil und das Ganze* (R. Piper & Co. Verlag, Munich).

Heisenberg, W. (1972). *La partie et le tout* (Albin Michel, Paris).

Heisenberg, W. (1989). *Theory, Criticism and Philosophy, in From a Life of Physics* (World Scientific, Singapore).

Hermann, A. (1971). *The Genesis of Quantum Theory (1899-1913)* (MIT Press, Cambridge – MA).

Hoffmann, D. (1984). *Erwin Schrödinger* (BSB B. G. Teubner Verlagsgesellschaft, Leipzig).

Van Hove, L. (1958). *Von Neumann's contributions to quantum theory.* Bulletin of the American Mathematical Society, **64**, pp. 95–99.

Klein, M. J. (1965). Einstein, Specific Heats and the Early Quantum Theory, *Science* **148**, pp. 173–180.

Klein, M. (Ed.) (1967). *Letters on Wave Mechanics* (Philosophical Library, New York).

Klein, M. J., Kox, A.J., Renn, J., and Schulman, R. (1994a) *The Collected Papers of Albert Einstein, Vol.3: The Swiss Years: Writings 1909-1911* (Princeton University Press).

Klein, M. J., Kox, A.J., Renn, J., and Schulman, R. (1994b) *The Collected Papers of Albert Einstein, Vol. 5: The Swiss Years: Correspondence, 1902-1914* (Princeton University Press, Princeton – NJ).

Buchwald, D. K., Illy, J., Kox, A.J., Lehmkuhl, D., Rosenkranz, Z. and James, J.N. (Eds.) (2021) *The Collected Papers of Albert Einstein, Vol. 15: The Berlin Years: Writings & Correspondence, June 1925–May 1927* (Princeton University Press).

Kox, A. J. (2009). *The Scientific Correspondence of H. A. Lorentz, Vol.1* (Springer, Berlin).

Kox, A. J. (2018). *The Scientific Correspondence of H. A. Lorentz, Vol.2* (Springer, Berlin).

Kragh, H. (2010). The Early Reception of Bohr's Atomic Theory (1913-1915): A Preliminary Investigation, *Research Publications and Science Studies* **9**, (Aarhus Centre for Science Studies, University of Aarhus), pp. 1–111.

Lambert, F. J. and Berends, F. (2021). *Einstein's Witches' Sabbath and the Early Solvay Councils. The Untold Story* (EDP Sciences, Les Ulis Cedex).

Langevin, A. (1966). Paul Langevin et les Congrès de Physique Solvay, *La Pensée, Revue du Rationalisme Moderne*, **129**, pp. 3–32.

Lettevall, R., Somsen, G., and Widmalm, S. (2017). *Neutrality in Twentieth-Century Europe Intersections of Science, Culture, and Politics after the First World War* (Routledge, Milton Park – UK).

Lorentz, H. A. (1934-1939). *Collected Papers*, Zeeman, P. and Ehrenfest, P. (Eds.) in 9 Volumes (Martinus Nijhoff, The Hague).

MacCormmach, R. (1966). The Atomic Theory of John William Nicholson, *Archive for History of Exact Sciences*, **3**, pp. 160–184.

MacCormmach, R. (1967). Henri Poincaré and the Quantum Theory, *ISIS* **58**, pp. 37–55.

Marage, P. and Wallenborn, G. (Eds.) (1999). *The Solvay Councils and the Birth of Modern Physics*, (Birkhaüser, Basel).

Marsden, E. (1954). The Rutherford Memorial Lecture, 1954. Rutherford – His Life and Work, 1871-1937, *Proceedings of the Royal Society of London A* **226**, pp. 283–305.

Mehra, J. (1975). *The Solvay Conferences on Physics: Aspects of the Development of Physics since 1911* (Springer Dordrecht, Berlin).

Mehra, J. and Rechenberg, H. (1987) *The Historical Development of Quantum Theory. Vol. 5. Erwin Schrödinger and the Rise of Wave Mechanics Part 2.* (Springer-Verlag, Berlin).

Michel, A. (Ed.) (1953). *Louis de Broglie, physicien et penseur* (Albin Michel, Paris).

Moseley, H. (1914). The High Frequency Spectrum of the Elements. Part II, *The Philosophical Magazine* **27**, pp. 703–713.

Navarro, J. (Ed.). (2018). *Ether and Modernity.The recalcitrance of an epistemic object in the early twentieth century* (Oxford University Press, Oxford – UK).

Owen, E.A. (1918). *Proceedings of the Royal Society of London* **94**, pp. 510–524.

Pais, A. *Niels Bohr Times: In Physics, Philosophy and Polity* (Clarendon Press, Oxford – UK).

Pelseneer, J. (unpublished manuscript). *Historique des Instituts de physique et de chimie Solvay, depuis leur fondation jusqu'à la Deuxième Guerre Mondiale.* A copy of the unpublished manuscript is part of the Archives of the Solvay Institutes, Brussels, S.a.b.ULB. A microfilm version is kept in the Archive for History of Quantum Physics, Library of the American Philosophical Society, (126, 495, 510).

Poincaré, H. (1911). Sur la théorie des quanta, *Comptes-rendus de l'Académie des Sciences*, **153**, pp. 1103–1108.

Poincaré, H. (1912). Sur la théorie des quanta, *Journal de Physique Theorique et Applquée*, **2**, pp. 5–34.

Poincaré, H. (1963). *Mathematics and Science: Last Essays (Dernieres Pensees). Translated from the French by John W. Bolduc* (Dover Publications, Inc.) – Originally published in French under the title of *Dernieres Pensees*, copyright (P) 1913 by Ernest Flammarion.

Rocci, A. and Lambert, F. J. (2024). The Solvay Councils, de Broglie's brothers, and the development of wave-particle duality, *Annales de la Fondation Louis de Broglie* **48-1**, pp. 45–90.

Rosenfeld, L. (1936). La première phase de l'évolution de la théorie des quanta, *Osiris* **2**, pp. 149–196.

Rotblat, J. (1945). Prof. Ludwik Wertenstein, *Nature* **156**, pp. 384–386.

Rutherford, E. (1911). The scattering of α and β particles by matter and the structure of the atom, *Philosophical Magazine*, **21**(125), pp. 669—688.

Rutherford, E. (1913). A New Physics Institute, *Nature* **90**, pp. 545–546.

Rutherford, E. (1913). The Structure of the Atom. *Nature* **92**, p. 423.

Schrödinger, E. (1924). Gasentartung und frei Weglänge, *Physikalische Zeitschrift* **25**, pp. 41–45.

Schrödinger, E. (1926), Quantisierung als Eigenwertproblem. Erste Mitteilung. *Annalen der Physik*, **384**, pp. 361–376.

Schrödinger, E. (1926), Quantisierung als Eigenwertproblem. Zweite Mitteilung. *Annalen der Physik*, **384**, pp. 489–527.

Schrödinger, E. (1926), Über das Verhältnis der Heisenberg-Born-Jordanschen Quantenmechanik zu der meinem. *Annalen der Physik* **79**, pp. 734–756.

Schweidler, E. (1906). Über Schwankungen der radioaktiven Umwandlung. In: Dunod, H. Pinat, E. (Eds), *Premier Congrès international pour l'étude de la radiologie et de l'ionisation tenu à Liège du 12 au 14 Septembre 1905*, (Imprimerie Scientifique Severryns, Bruxelles).

Slater, J.C. (1975). *Solid state and molecular theory: A scientific biography* (Wiley, New York).

Soddy, F. (1914). *The chemistry of the radio-elements, Part II*, (Longmans, Green & Co., London).

Solvay I – Solvay, E., Langevin, P., Broglie, M. D. D., & Institut international de physique Solvay (1912). *La théorie du rayonnement et les quanta: rapports et discussions de la réunion tenue à Bruxelles du 30 octobre au 3 novembre 1911.* (Gauthier-Villars, Paris).

Solvay II – Institut international de physique Solvay. (1921). *La structure de la matière: rapports et discussions du Conseil de physique tenu à Bruxelles du 27 au 31 octobre 1913* (Gauthier-Villars, Paris).

Solvay III – Institut international de physique Solvay. (1923). *Atomes et électrons: rapports et discussions du Conseil de physique tenu à Bruxelles du 1er au 6 avril 1921* (Gauthier-Villars, Paris).

Solvay IV – Institut international de physique Solvay. (1927). *Conductibilité électrique des métaux et problèmes connexes: rapports et discussions du quatrième Conseil de physique tenu à Bruxelles du 24 au 29 avril 1924* (Gauthier-Villars, Paris).

Solvay V – Institut international de physique Solvay. (1928). *Electrons et photons: rapports et discussions du cinquième Conseil de physique tenu à Bruxelles du 24 au 29 octobre 1927* (Gauthier-Villars et cie).

Solvay VI – Institut international de physique Solvay. (1932). *Le magnétisme: rapports et discussions du sixième Conseil de physique, tenu à Bruxelles du 20 au 25 octobre 1930* (Gauthier-Villars, Paris).

Solvay XII – Institut international de physique Solvay (1962). *La théorie quantique des champs: douzième Conseil de physique, tenu à l'Université Libre de Bruxelles du 9 au 14 octobre 1961* (International Publishers, New York).

Solvay XXIII – Gross, D., Henneaux, M., and Sevrin, A. (Eds.). (2007). *The Quantum Structure of Space and Time. Proceedings of the 23rd Solvay Conference on Physics Brussels, Belgium 1 – 3 December 2005.* (World Scientific, Singapore).

Solvay XV – Gross, D., Henneaux, M., and Sevrin, A. (Eds.), *The Theory of the Quantum World. Proceedings of the 25th Solvay Conference on Physics (19—22 October 2011)*, (World Scientific, Singapore).

Stark, J. (1913). Beobachtungen über den Effect des elektrischen Feldes auf Spectrallinien. Vorgelegt von Rubens, *Sitzungsberichte der Königlich preußischen Akademie der Wissenschaften*, Sitzung der 20 Nov. 1913.

Steward, E. G. (2008). *Einstein's Transition Probabilities: Bohr's theory and Planck's Law* (World Scientific, Singapore).

The Nobel Foundation (ed.) (1962). *Nobel. The Man and his Prizes* (Elsevier Publishing Company, Amsterdam).

Thomson, J. J. (1908). *On the light thrown by recent investigations on electricity on the relation between matter and ether : the Adamson Lecture delivered at the University on November 4, 1907.* (Manchester University Press, Manchester).

Trillat, J. (1960). Maurice de Broglie (1875-1960), *Bulletin de la Société française de Minéralogie et Cristallographie,* **83**(10-12), pp. 238–241.

Uhlenbeck, G. E. and Goudsmit, S. (1925). *Natürwissenschaften* **13**, pp. 953–954.

Wheaton, B. (1983). *The Tiger and the Shark. Empirical Roots of Wave-Particle Dualism* (Cambridge University Press, Cambridge – UK).

Wien, W. (1905). Über die Energie der Kathodenstrahlen im Verhältnis zur Energie der R—"ontgen- Sekundärstrahlen, *Annalen der Physik,* **18** (Iss. 5), pp. 991–1007.

Wright, J. R. (1911). Das positive Potential des Aluminiums als eine Funktion der Wellenlänge des einfallenden Lichtes, *Physikalische Zeitschrift,* **12**, pp. 338–343.

Index